GCSE Technology

A021865043

D1796163

GCSE TECHNOLOGY

Steve Rich   Tony Edwards

Other titles in this series:

PNEUMATICS

Steve Rich

Head of Design, Weatherhead High School,
Wirral, Merseyside

Anthony Edwards

Head of Design Technology
at a Wirral Sixth Form College, Merseyside

Stanley Thornes (Publishers) Ltd

First published in 1990 by:
Stanley Thornes (Publishers) Ltd
Old Station Drive
Leckhampton
CHELTENHAM GL53 0DN
England

British Library Cataloguing in Publication Data

Rich, Steve
Pneumatics.
1. Pneumatic engineering
I. Title   II. Edwards, Anthony   III. Series
621.51

ISBN 0-85950-933-8

Acknowledgements

The authors and publishers are grateful to the following for providing photographs for this book:

Ann Ronan Picture Library page 79 (bottom); Charles Tait Photographic page 80 (top); Compair Maxam pages 10, 12, 86; Eyeline Photos page 79 (top); Hydrovane page 8; Hydrovane/Economatics Education page 9 (top); Philip Harris Education page 9 (bottom); Science Photo Library page 80 (bottom).

All other photographs were taken of Testbed Technology Equipment by Anthony Price.

Diagrammatic artwork by Mark Dunn
Typeset in 11/12½ Italia by Tech-Set, Gateshead, Tyne & Wear
Printed and bound in Great Britain at The Bath Press, Avon

Contents

Preface

This book is designed to be a self-contained course that will be 'user friendly' to both student and teacher. It contains a series of graded design problems that can be used sequentially or independently. This allows flexibility in the use of the material.

The book takes into account the varied requirements of the National Curriculum through the medium of high technology. It has been tried and tested as teaching and reference material for students preparing for GCSE technology examinations (equivalent to Key Stage 3 upwards).

The authors recognise that not all schools have the same level of equipment and therefore the development of a specification for each design brief has been omitted. It is envisaged that this can be developed by the students who need to be aware of the equipment available to them.

Sample calculations and relevant equations have been included in Chapter 25. It is intended that more able pupils can use this information to develop extension studies based on the context of the design. For the less able, teachers could introduce and explain the calculations when and if they are required.

Chapter 26 catalogues common types of pneumatic components. It is designed to be used in the same way as Chapter 25, as a basis for research and extension studies.

All diagrams in this book have been drawn to British Standards (reference BS 2917). All circuits have been thoroughly tested with a variety of systems.

Steve Rich   Anthony Edwards
1990

CHAPTER 1

Students' guide to the book

Every chapter in this book contains information on a particular aspect of pneumatics together with a theoretical and/or practical activity to help you to develop your understanding. Each new section builds on the work done earlier in the book. There are additional activities in the teacher's book for this series.

When doing practical work you should be aware that incorrect use of compressed air and pneumatic devices can be dangerous. We recommended that you read Chapter 5 *Safety* before starting your practical work with pneumatic components.

As you read through the book you will find that each section can be used on its own to explore a particular topic or idea or it can be combined with the others to give you a complete course in pneumatics.

To avoid confusing detail the *Analysis/Research* sections do not include all the work needed to fully design and install a complete solution to the design problems. You will need to find more information if you are to model the solutions accurately.

You may not need to read all the chapters. Seek your teacher's advice on which sections of this book are most important in terms of your examination syllabus. If you do complete every chapter your project work will be helped by having a wider understanding of pneumatic systems.

Good luck!

1

Everyone uses compressed air

A simple activity like blowing up a balloon involves the use of **compressed air**. Footballs and netballs are inflated by the same process.

Everyone uses compressed air to do work

When you pump up the tyres on a bicycle they are inflated by compressed air. The energy generated by the pumping action is stored in the air in the tyre. This air holds the wheel rim away from the ground. It helps to cushion the rider from the small bumps in the road.

You will already have experienced some of the uses of compressed air and the work that it can do in everyday life.

 ASSIGNMENT 2.1

Here are some illustrations of different devices. List the ones you think are pneumatically operated. State what the energy from the compressed air is converted into in each device.

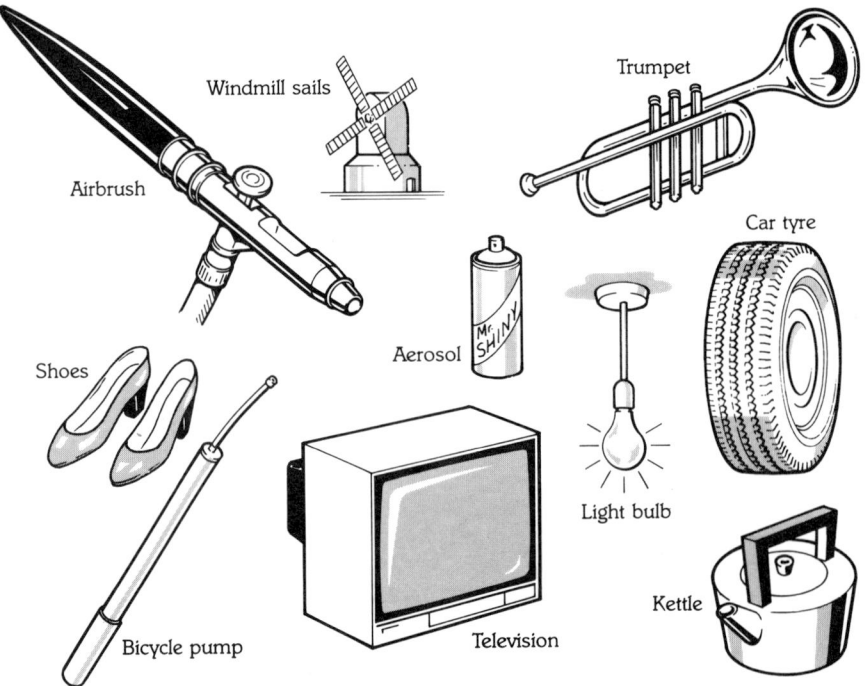

Airbrush

Windmill sails

Trumpet

Car tyre

Aerosol

Shoes

Light bulb

Bicycle pump

Television

Kettle

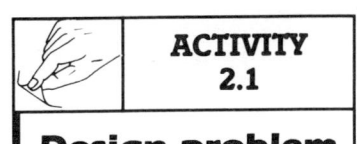

ACTIVITY 2.1	A younger member of your family would like to make a new toy. There are plenty of raw materials available and you have been asked to help.
Design problem	

Design brief

Using items such as those listed below, design and make a simple, pneumatically powered toy that will amuse a young member of your family.

Analysis/Research

This section contains information that you need to consider when designing your solution to a problem.

Children like toys that move.

You may find some useful ideas in the list of devices that you produced for Assignment 2.1.

Suitable materials

Balloons	Pins	Washing-up/fizzy drink bottles
Thread or string	Card	Drinking straws
Balsa wood	Paper	Cardboard or wooden discs
Clingfilm	Plastic tubing	Dowel rod

Useful tools

Adhesive tape	PVC insulating tape	Glue gun and glue
Paper/card glues	Scissors	Modelling knife
Hand drill and drills	Junior hacksaw	

Evaluation

When you have designed, assembled and tested your device, explain what were its good points and what things you would change if you made it again.

CHAPTER 3

Pneumatics and motion

From the examples in Assignment 2.1 it is clear that different pneumatic devices are designed to produce different types of movement. The compressed air in a dentist's drill forces the cutting bit to rotate. The compressed air used in a tube train opens and closes the doors.

We need to understand four main types of motion in order to fully describe any pneumatic device. To help us work out what happens in a machine we need to look at the **input** motion to it and the resulting **output** motion from it.

We shall look at a dentist's drill in detail.

The dentist operates the drill by pressing a foot pedal (**input**). The pedal **controls** the flow of air through a regulating valve. The drill is caused to revolve by the flow of air (**output**).

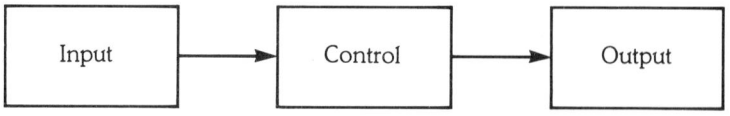

The input and output are different forms of motion and we need accurate words to describe them.

Linear motion

This is one-way movement along a straight line. It is represented by an arrow like this:

Reciprocating motion

This is movement up and down, left and right or forwards and backwards along a straight line. It is represented by arrows like this:

up and down, or

left and right, or

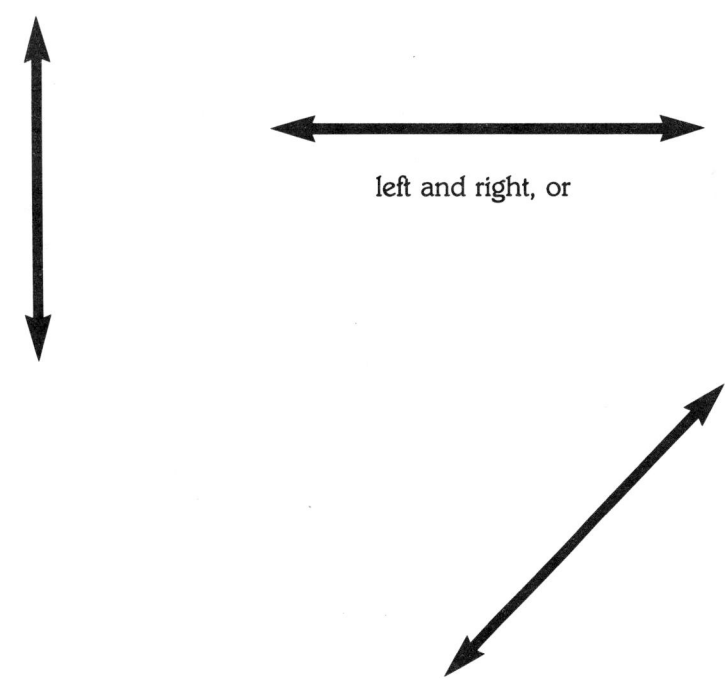

forwards and backwards along a straight line

The speed of the reciprocating motion is not constant. It changes all the time, accelerating and decelerating as the direction changes.

Oscillating motion

This is the sort of movement the pendulum on a clock makes. It is like reciprocating motion, but along an arc.

This kind of motion is represented by a curved line with an arrowhead on both ends, like this:

Rotary motion

This is motion in a circular direction.

It is represented by a curve with an arrowhead indicating the direction of movement, like this:

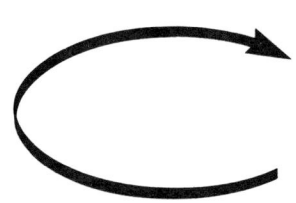

ASSIGNMENT 3.1

Try to identify the input and output motions for each of the devices listed on page 7. Use the graphical symbols already described to communicate and explain your answers. The first one has been done for you.

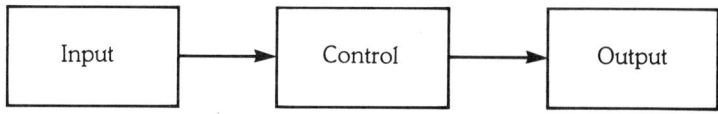

Input → Control → Output

a) Dentist's drill

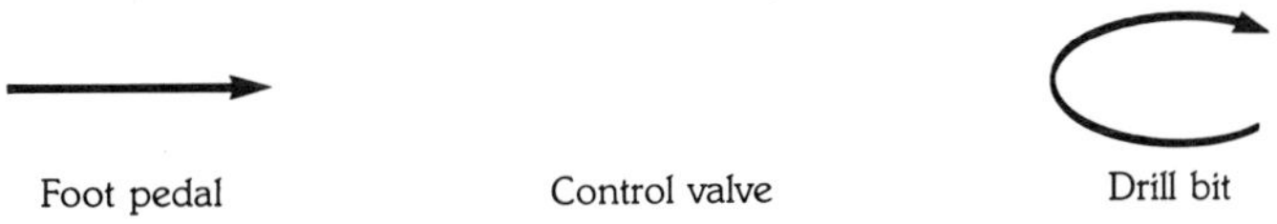

⟶

Foot pedal   Control valve   Drill bit

b) Pneumatic drill
c) Perfume atomiser
d) Bicycle pump
e) Windsurfer sail
f) Trumpet
g) Blowpipe
h) Windmill sails
i) Aerosol
j) Car tyre
k) Airbrush

Supplying compressed air

Compressed air can be produced by a device like a bicycle pump. Simple pumps operated by people can generate small volumes of compressed air but this involves hard work. The energy required to compress large volumes of air is sometimes beyond the ability of the human frame.

In order to do large amounts of work with pneumatic devices powerful machines are needed to compress the air. A pneumatic drill is a common sight. To compress the air for the drill a portable diesel-driven **compressor** is used. It is a noisy device.

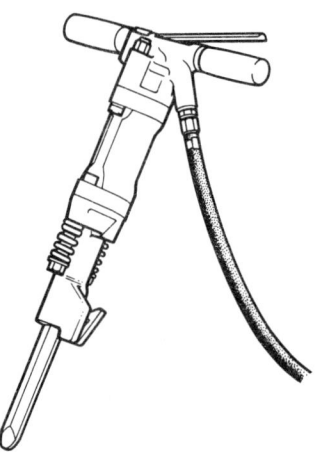

The compressor and the **motor** that powers it are linked to form one unit. This unit converts the chemical or electrical energy (which drives the motor) into potential energy, stored as compressed air.

Compressor

In a working environment large, noisy compressors are unacceptable. Small electrically-driven compressors are available that produce much less disturbance. These fall into two types; those with a **receiver** and those without. A receiver is a container which stores compressed air.

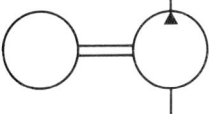

Symbol

Small electrically driven motor/compressor unit

Direct-supply compressors that do not have a receiver must run continuously to give a constant supply of compressed air.

The units that have a motor, compressor and receiver as one unit do not have to run all the time once the receiver is full. Air can be supplied from the charged receiver.

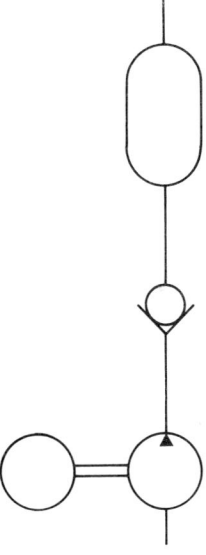

Symbol

Motor/compressor/receiver unit. The drain valve is at the bottom of the receiver.

Every compressor unit should have a cutout that will turn it off when the pressure in the system reaches a preset level. This job is done by a **pressure relief valve**. If the pressure is too high the valve opens and allows the compressed air to escape into the atmosphere.

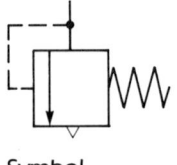

Symbol

Pressure relief valve

When air is compressed in the compressor it becomes hot. As soon as the air leaves the compressor it cools. Any moisture in it condenses into droplets of water. This water can damage pneumatic components by causing corrosion. Receiver units are fitted with a **drain valve** to allow the water to be removed.

A **shut-off valve** is placed on the output from the compressor to prevent air escaping into the system when it is not required.

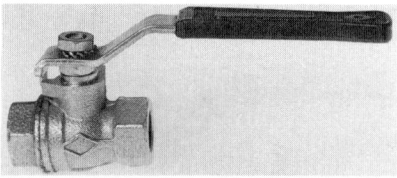

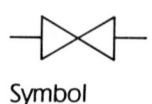

Symbol

Shut-off valve

To ensure that the air leaving the compressor is clean and dry a **filter** is connected directly to the output from the receiver. This removes water and dust from the air as it flows through.

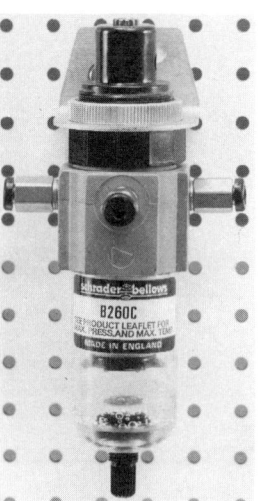

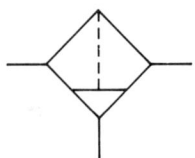

Symbol

Filter

The compressed air then passes through a **regulator** that allows the output pressure to be controlled. It also enables the output pressure to remain constant despite changes in pressure in the receiver. The regulator should be fitted with a pressure gauge to give a reading of the output.

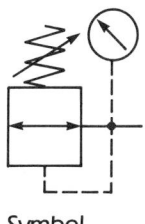

Symbol

Pressure regulator

Air pressure can be measured in **pascals**. A pascal is equivalent to one newton of force per square metre. This is a very small unit and for practical purposes the **bar** is used.

$$1 \text{ bar} = 100\,000 \text{ pascals}$$

(One bar is the same as the atmospheric pressure we experience at sea level.)

In some pneumatic systems an **oil lubricator** is added. This ensures that the insides of all the valves are lubricated to protect them from corrosion and to help them work efficiently. This is not essential for school systems but does help to prolong the life of the components.

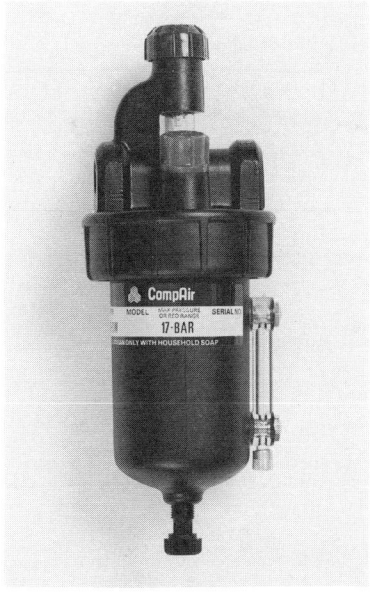

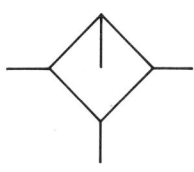

Symbol

Oil lubricator unit

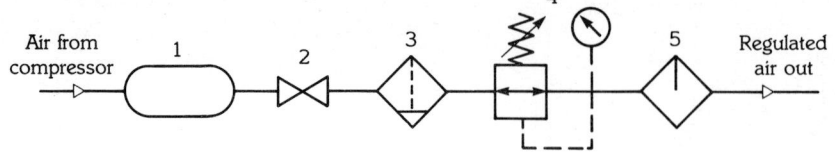

1) Identify the components in the diagram above. Copy the diagram and explain:
   a) What the components are.
   b) What they do in this circuit.
   The components should be connected in the order shown above.

2) Using the following list for reference look at your own school's pneumatic system and identify which components are used in it.

   Receiver
   Shut-off valve
   Pressure relief valve
   Air filter
   Pressure regulator
   Pressure gauge
   Oil lubricator
   Drain valve

CHAPTER 5

Safety

In a controlled situation pneumatic devices and systems are safe. If they are used incorrectly they can present serious hazards to the user.

To ensure that you work safely, here are some simple guidelines.

1) Never point a live airline at yourself or anyone else.
Compressed air is very powerful and can cause severe damage to the skin.

2) Protect your eyes with goggles when doing practical work.
A compressed air jet pointed at the eyes can cause a very serious injury. Loose particles on the worktop can also be blown into the face by exhaust air from cylinders.

3) Cover any open cuts on your skin with a plaster.
Compressed air can be forced through an open cut into the blood vessels and cause a similar illness to the 'bends' (as suffered by divers).

4) Never try to stop components in motion with your hands.
Because compressed air can store a great deal of energy, pneumatic components can have very powerful movements.

5) Make sure that all pipework is secure.
Loose pipes can whiplash violently if air is passed through them.

6) Always consult your teacher to check that your circuit is safe.

Pneumatic systems are widely used in the world outside school. They provide a safe and powerful way of controlling movement.

Design problem

During a group project, pupils have designed a new frame for a range of seats. They now want to test them. Having made sample frames in both wood and metal they want to know which is more durable.

Design brief

Design a simple testing device that will simulate normal conditions of use. The shape of the frame is shown below. The load is to be applied at the marked point, half-way along the upper arm.

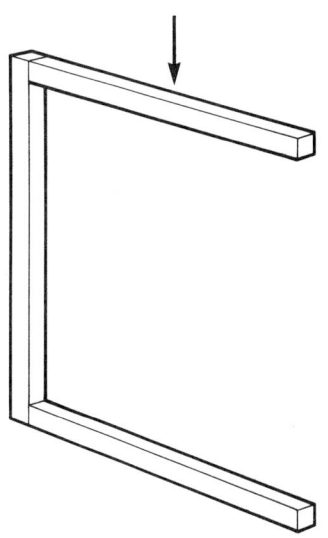

Analysis/research

A pneumatic solution is appropriate because:

a) The forces applied to the arm can be easily calculated.
b) It is simple to operate and maintain.
c) There is already an existing pneumatic system available.
d) A pneumatic solution does not require a large amount of space.

The testing load is applied on the downward stroke of the device.

A reciprocating action is required by the testing device to simulate real conditions.

A common pneumatic device used to produce a reciprocating output motion is a **single-acting cylinder**.

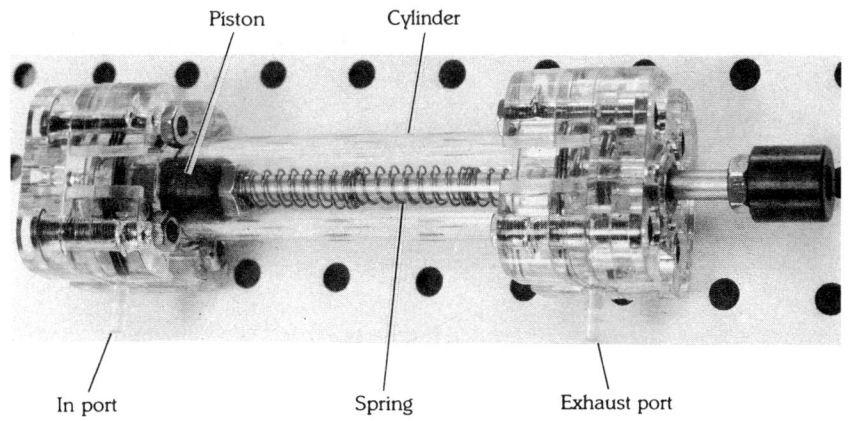

Single-acting cylinder

When compressed air enters the cylinder it forces the **piston** to move on an outwards stroke.

A **compression spring** is put into the cylinder which pushes the piston back to its original position when the compressed air is turned off.

When this type of cylinder is drawn in a circuit diagram this symbol is used:

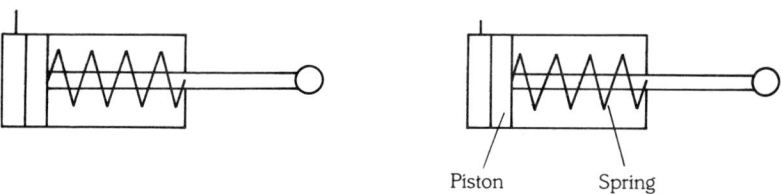

Piston   Spring

When the piston is sent out on its power stroke it is said to be going **positive**.

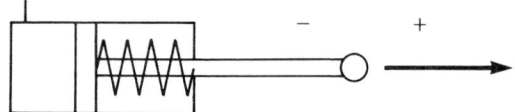

When the air pressure is removed the spring pushes the piston back to where it started. In this situation the piston is said to be going **negative**.

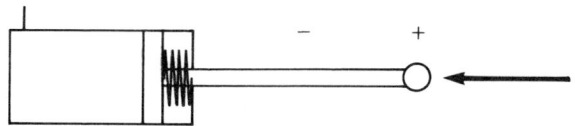

This cylinder is called a **single-acting cylinder** because it can only *act powerfully* in a *single* direction.

Note: Single-acting cylinders can be purchased with the spring on the opposite side of the piston to give an automatic positive stroke and a working negative stroke.

A method of controlling the single-acting cylinder is required. (This research continues in the next chapter.)

Three-port valves

In a human body **valves** are used to control the flow of blood in the heart, arteries and veins. In a similar way different types of valves are used to regulate the flow of compressed air in a pneumatic system.

Three-port valves are used to control the flow of air in a circuit. The **ports** of the valve can be opened or closed by a mechanical, electrical or pneumatic switch.

One design of a **push-button spring-return three-port valve** is shown below. In normal use port 1 is connected to the air supply, port 2 is connected to the cylinder and port 3 is left unattached to allow exhaust air to escape, as shown in the simplified diagrams.

OFF position

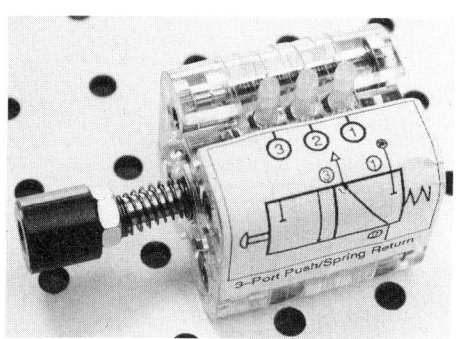
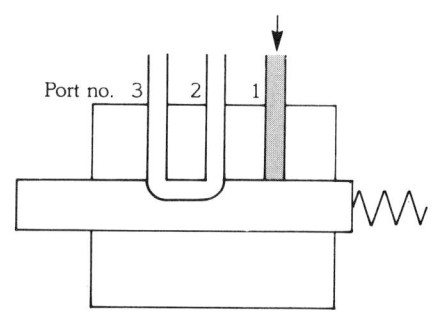

In the OFF state (not pressed) port 1 is closed and compressed air cannot pass through it.

ON position

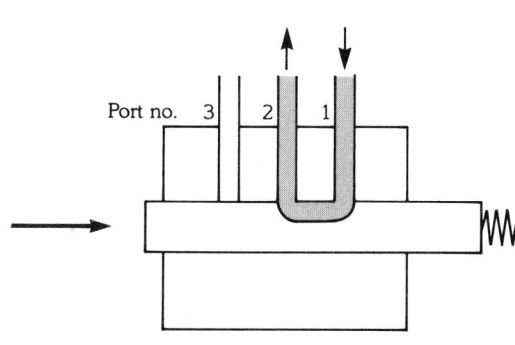

When the valve is operated (pressed) it allows air to pass in through port 1 and out through port 2.

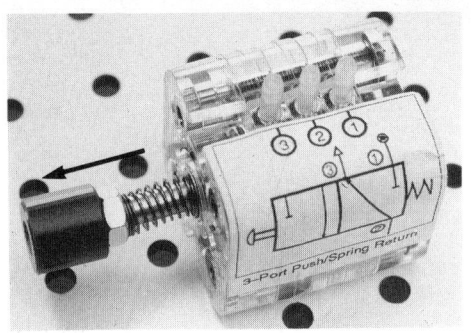

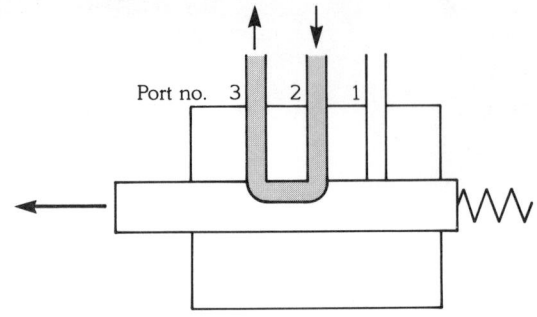

Port no. 3 2 1

When the push-button is released the valve returns to the OFF position. Used air can now flow into port 2 and escape into the atmosphere from port 3.

Remember: Compressed air is still connected to port 1.

ASSIGNMENT 7.1

Make a list of five or more devices that use valves to regulate or control their performance.

Extension
Explain in detail how one of the devices in your list is controlled by the valves.

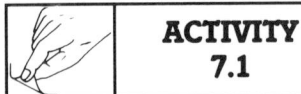

ACTIVITY 7.1

Using some or all of the components listed below design and make a model to illustrate how a three-port valve works.

T connector
Plastic or rubber tubing
Balloon
Thread
Rubber bands

You can use your hands as part of the mechanism.

(This research continues in the next chapter.)

Circuit diagrams for three-port valves

Pneumatic circuits can be assembled with components made by many different manufacturers. The valves all look different.

These are both push-button spring-return three-port valves

To make designing and drawing circuits simpler standard symbols and diagrams are used.

The circuit symbol for a three-port valve has to show several important features:

a) The ON position
b) The OFF position
c) The type of switch used on the valve.

The valve is drawn as two linked boxes. The top box shows the ON state and the lower box the OFF state. The three connections are drawn on the sides of each box.

In the OFF position ports 2 and 3 are connected. This is shown as:

The empty arrow head connected to port 3 shows that the air can escape from this port to the atmosphere.

The T shape shows that port 1 is closed inside the valve.

The black arrow head shows the direction in which the air is flowing through the valve.

In the ON position ports 1 and 2 are connected. This is shown like this:

If the valve is operated by a push-button it is drawn as:

The compressed air supply is shown as: ⊙—

When the valve is released a spring pushes the valve back to the OFF position. The spring is shown as: ⧚

The complete valve symbol looks like this:

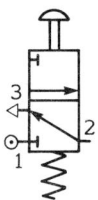

This circuit diagram shows the valve in its OFF state.

The differences between the valve in the ON and the OFF state are shown below. Compare the two drawings.

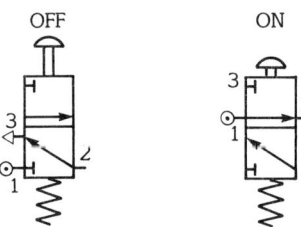

Note: The ON position box is at the top of the diagram.
The OFF position box is at the bottom of the diagram.

The valve can be drawn turned to the left or right:

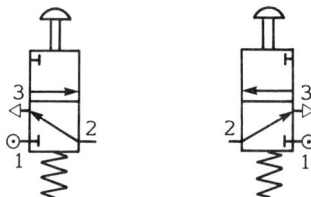

The symbols for all pneumatic components must conform to the **BSI/ISO** (British Standards Institute/International Organisation for Standardisation) standards.

Design solution

By combining all the information gained in the research in Chapters 6 and 7 and this chapter, a solution to the design problem in Chapter 6 can be proposed:

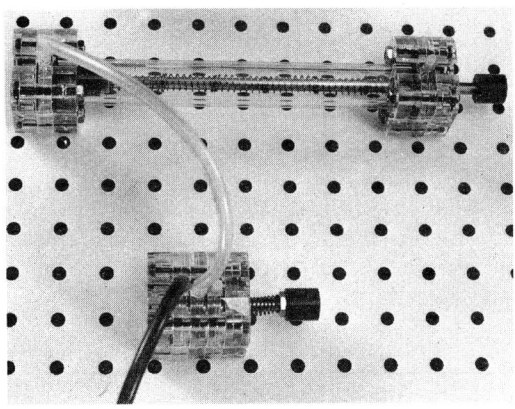

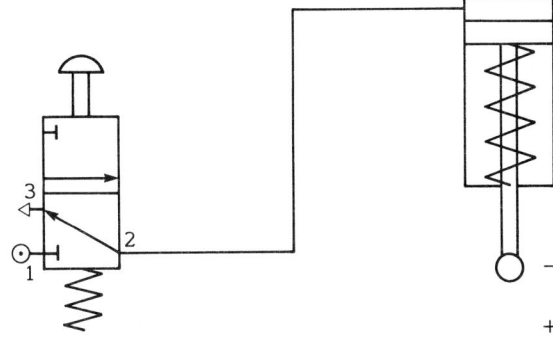

Evaluation

A push-button spring-return three-port valve connected to a single-acting cylinder will give the required operation. The piston goes positive when the three-port valve is pressed and returns negative when the valve is released.

Note: In this type of circuit the positive stroke is the most powerful. The negative stroke is only powered by a spring.

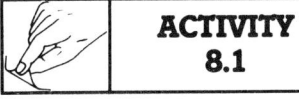

ACTIVITY 8.1

Assemble and test the circuit shown above. Copy the circuit diagram. Test your circuit. Note the powerful movement of the piston. Can you think of another use for this system?

OR gates

Design problem

A hypermarket has a pneumatically operated sliding door on its deep freeze room. The door needs to be kept closed as much as possible to maintain a constant low temperature. A large single-acting cylinder had been installed to move the door. The spring is sufficiently strong to close the door slowly.

The door also needs to be opened by workers inside and outside the cold room.

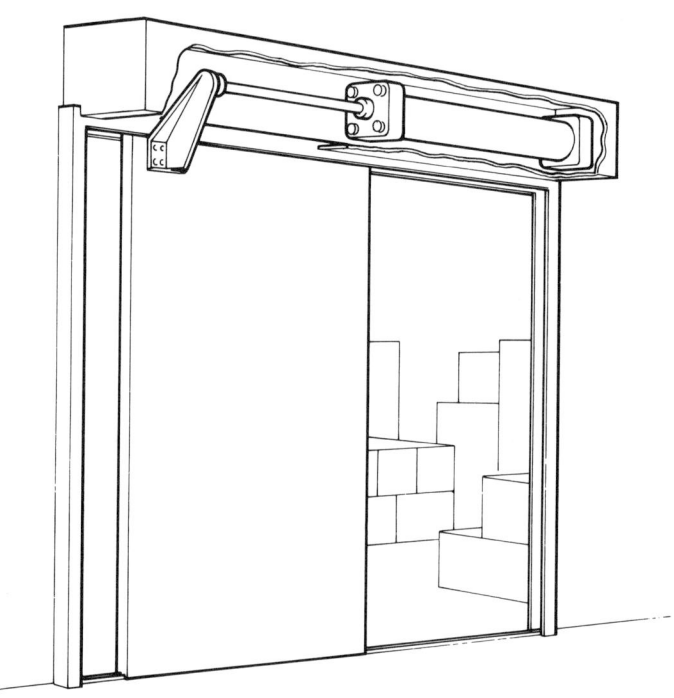

Design brief

Design a pneumatic control system that will allow the cold-room door to be opened from either side.

Analysis/Research

Two valves are required, one inside and one outside the room.

They should be suitable for operation by either hand or foot.

Both valves need to be connected to the port on the single-acting cylinder by suitable pipework.

The pipes from the two valves need to be joined together, probably before they are connected to the cylinder.

Design solution 1

This design has two push-button spring-return three-port valves connected to a single-acting cylinder.

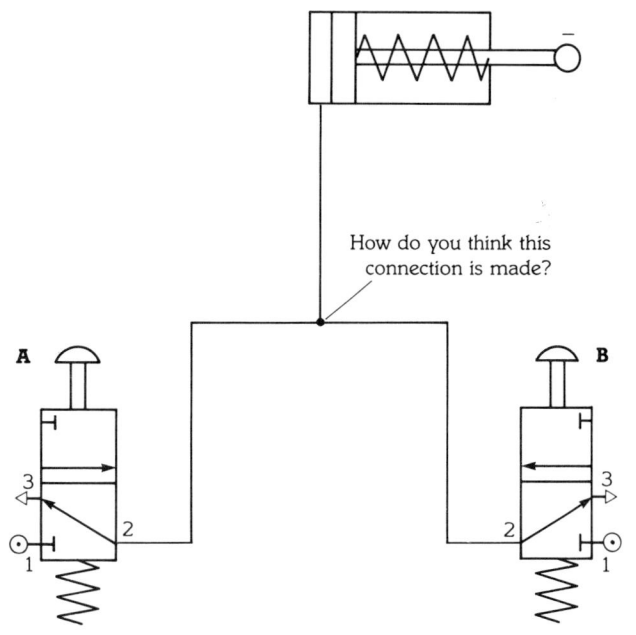

How do you think this connection is made?

A   B

Evaluation

When this design was assembled and tested it was found that the single-acting cylinder worked when either valve **A** or valve **B** was pressed. There was a major problem, however. Air was able to escape from port 3 on the valve not being pressed. This made the system very noisy and inefficient.

New design problem

A method must be found of modifying the circuit above to stop the air escaping from the unused valve.

Research

Shuttle valve

A suitable component that could be used is a **shuttle valve**. It is used to direct a signal to a cylinder from one of two valves.

This is the symbol for a shuttle valve:

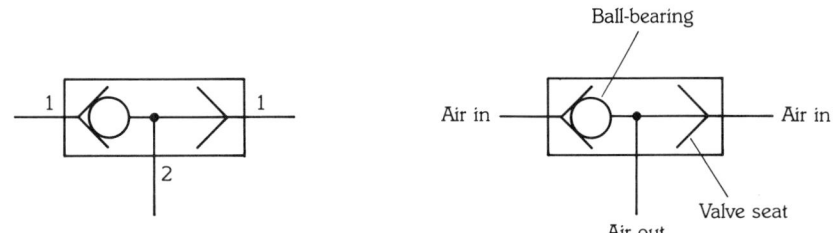

When air arrives at the left-hand side it enters the valve through port 1. This causes a small ball-bearing to move and block the right-hand port. The air now leaves the valve through port 2. When air enters from the right-hand side, the ball-bearing moves across the other way and closes the left-hand port. The air again leaves from port 2.

Design solution 2

The new circuit has a shuttle valve replacing the T connector.

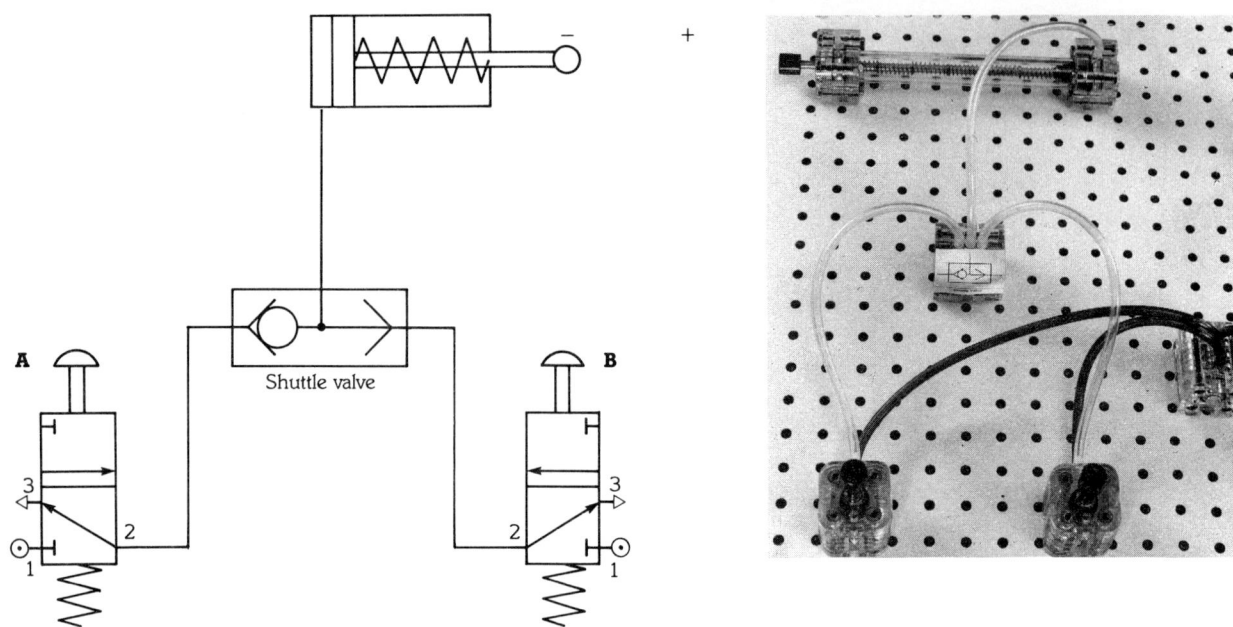

Shuttle valve

Evaluation

When either valve **A** or valve **B** is operated the piston is sent positive and the door is opened. When the valve is released the spring inside the cylinder causes it to go negative and shut the door.

This arrangement of valves is called an **OR gate**, because either valve **A OR** valve **B** may be operated to send the cylinder positive.

In the environment where this circuit will be used, most of the people using the door will be carrying something. Is there a more suitable type of switch that could be used?

ACTIVITY 9.1

Collect:
2 three-port valves (they need not be push-button)
1 single-acting cylinder
1 shuttle valve
Connecting pipes

Connect them together to make the circuit above. Test it.

ASSIGNMENT 9.1

1) Copy and complete this **truth table**. (The first line has been done for you.)

Valve A	Valve B	Cylinder position
OFF	OFF	Negative
ON	OFF	
OFF	ON	
ON	ON	

2) Add a third three-port valve **C** in the position shown below.

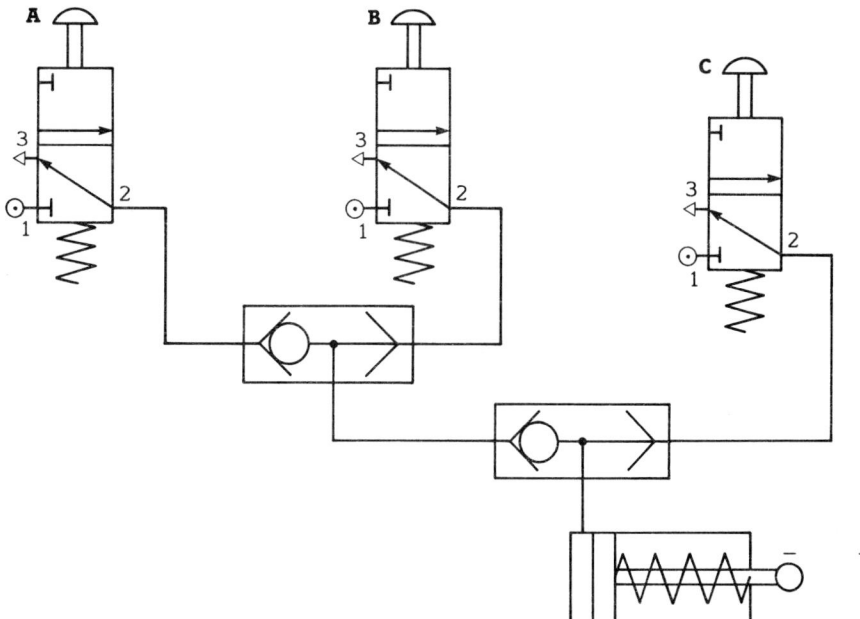

Operate the circuit again.

a) Does the circuit still work as it did before? Explain what happens.

b) Copy and complete this new truth table:

Valve A	Valve B	Valve C	Cylinder position
OFF	OFF	OFF	Negative
OFF	OFF	ON	
OFF	ON	OFF	
OFF	ON	ON	
ON	OFF	OFF	
ON	ON	OFF	
ON	OFF	ON	
ON	ON	ON	

CHAPTER 10

AND gates

With pneumatically operated machines it is important to protect the operators from injury. Powerful cylinders can badly damage hands or fingers if they get caught in the machine. These machines must be fitted with **guards** that enable them to be operated safely.

Design problem

A pneumatically powered press is used to make food containers out of aluminium foil for take-away restaurants.

Health and safety regulations require the machine to be fitted with an efficient guarding system. It has been decided to link the guard to the operation of the machine. The press can only operate when the guard is held in place.

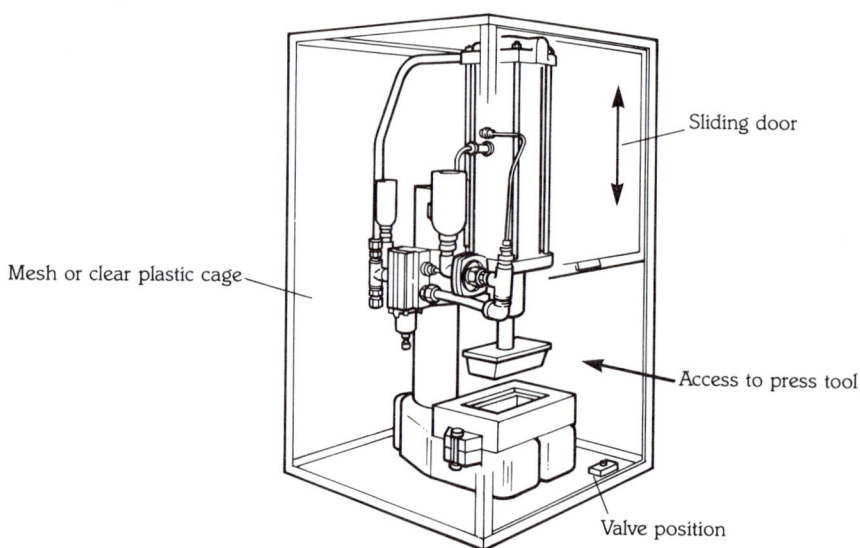

Sliding door

Mesh or clear plastic cage

Access to press tool

Valve position

Design brief

Design a pneumatic control system that will allow the pressing machine to run only when the safety guard is correctly in place. The operator should have to use both hands to operate the machine.

Analysis/Research

A pneumatic solution is appropriate because the press is powered by compressed air.

A pneumatic start switch is already installed on the machine.

An additional switch will be needed to detect when the guard is correctly in place.

Both switches must be pressed to start the machine.

One valve should be suitable for operation by a hand. The other valve will need to be activated by the safety guard.

The switches must only stay on for as long as they are pressed – there must be no chance of an accidental operation of the cylinder.

Design solution 1

This design has a push-button spring-return three-port valve and a roller-trip spring-return three-port valve connected in series to a single-acting cylinder:

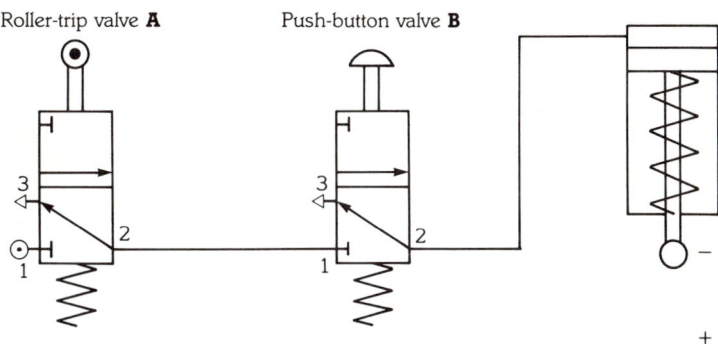

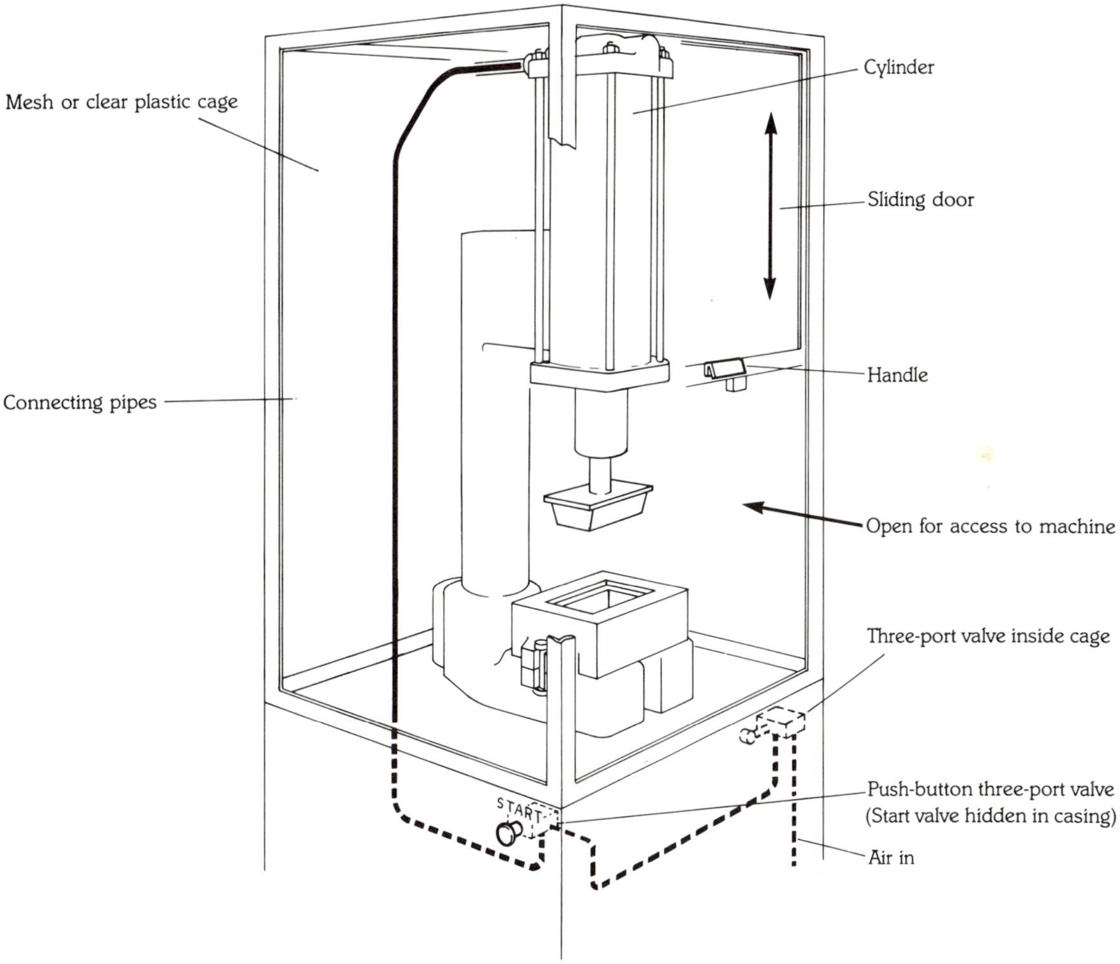

Roller-trip valve **A**   Push-button valve **B**

Mesh or clear plastic cage

Connecting pipes

Cylinder

Sliding door

Handle

Open for access to machine

Three-port valve inside cage

Push-button three-port valve (Start valve hidden in casing)

Air in

START

Evaluation

When assembled this circuit works well. The press (the single-acting cylinder) will only go positive when both valves are operated at the same time. In the industrial context, however, the position of the valves is important. The push-button valve must be located so that it can only be operated by a hand (not by a pile of books, a lump of metal, etc.).

During a production run the operator may hold the push-button valve down all the time and use one hand to remove the container, place the foil in the press and then close the guard. This is not a **fail safe** system because air will be present at the roller trip valve all the time. If the operator touches the roller trip whilst placing or removing the foil the press will operate. An accident may then result.

ACTIVITY 10.1

Design solution 1 shows a circuit diagram for a suitable pneumatic system. The operator must press both valves **A** and **B** before the cylinder will go positive.

Collect: 2 three-port valves (they need not be push-button)
            1 single-acting cylinder
            Connecting pipes

Connect them together to make the circuit. Test it.

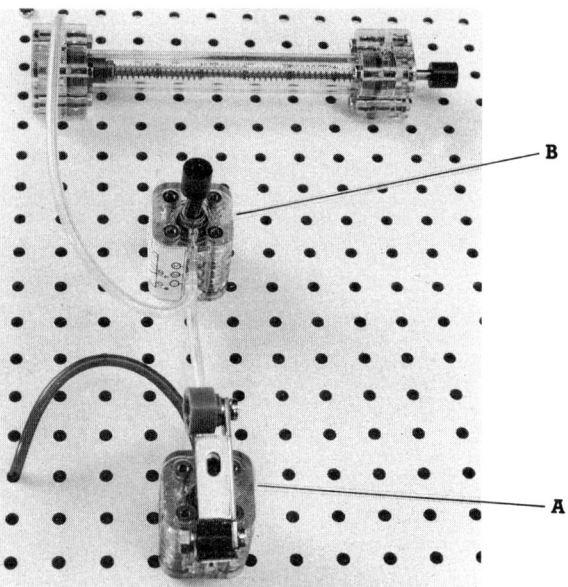

When valve **A** is pressed air travels down the pipe to valve **B**. When valve **B** is pressed the air is able to travel to the cylinder sending it positive. When either valve is released the cylinder returns negative.

The arrangement of valves is called an **AND gate**, because valve **A** **AND** valve **B** have to be operated together to send the cylinder positive.

ASSIGNMENT 10.1

1) Copy and complete this truth table for the circuit you have just assembled. The cylinder should start in a negative position. When the valves are operated the piston will either go positive or remain in a negative position.

Valve A	Valve B	Cylinder position
OFF	OFF	Negative
ON	OFF	
OFF	ON	
ON	ON	

2) Add a third three-port valve **C** in the position shown (see opposite page). (Note the reversed drawing of valve **A**; it makes no difference to the practical circuit.)

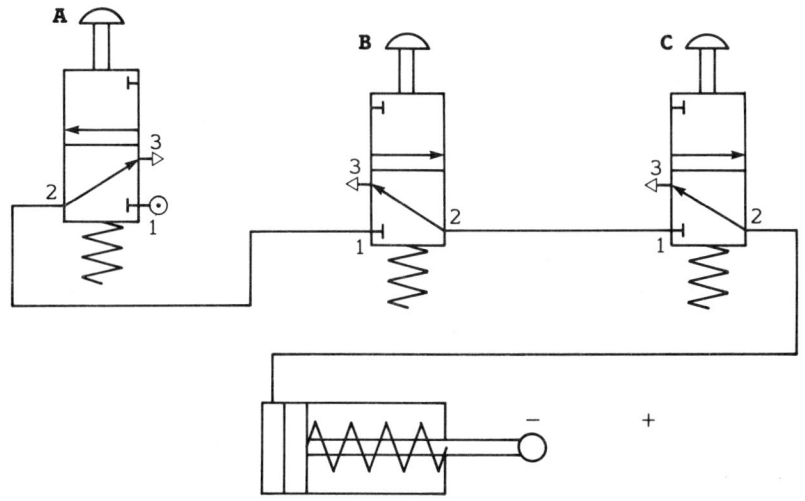

a) Operate the circuit again.
b) Complete this truth table:

Valve A	Valve B	Valve C	Cylinder position
OFF	OFF	OFF	Negative
OFF	OFF	ON	
OFF	ON	OFF	
OFF	ON	ON	
ON	OFF	OFF	
ON	ON	OFF	
ON	OFF	ON	
ON	ON	ON	

Can you think of an alternative situation where an AND gate circuit would be used?

NOT gates

Design problem

The entrance to a bank vault must be kept locked all the time except for a few brief seconds when the bank staff enter. The vault is permanently staffed to allow work to continue through the night.

Design brief

Design a simple circuit that will keep the door locked until the operator inside the vault presses a button to allow staff to enter or leave.

Analysis/Research

We can analyse this situation by completing the truth table:

Input	Output
No button pressed	Bolt positive
Button pressed	Bolt negative

In pneumatic terms,

Input	Output
Valve OFF	Cylinder positive
Valve ON	Cylinder negative

The system needs to remain positive until a valve is pressed.

Design solution

The normal connections for a three-port valve are shown on the left-hand diagram below. By changing the connections and direction of air flow the valve's operation can be changed.

Normal connections   NOT connections

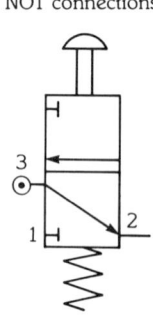

The diagram on the right (see opposite page) shows the valve connected as a **NOT gate**. The valve allows air through when it is **NOT** operated.

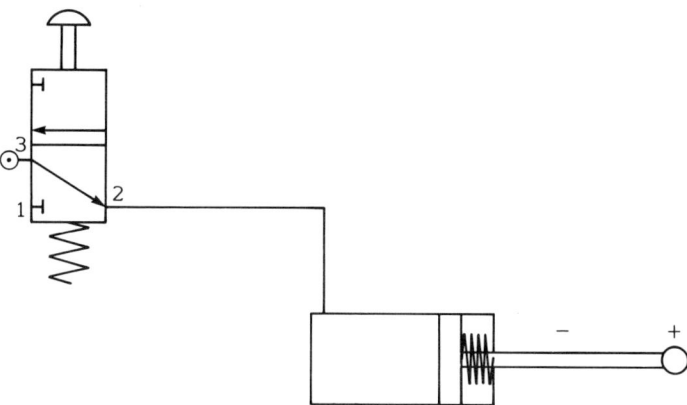

Evaluation

This circuit will work in practice. The cylinder will remain positive until the valve is operated. The cylinder will be sent negative when the valve is pressed. Because a single-acting cylinder is used the bolt will be held in its locked position under the action of compressed air. This has the advantage of keeping the lock secure.

Note: Some three-port valves leak badly if connected as NOT gates. Try valves made by different manufacturers if you have problems.

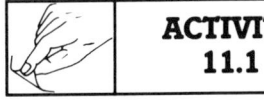

ACTIVITY 11.1

Collect: 1 three-port valve
1 single-acting cylinder
Connecting pipes

Connect them and make the circuit. Test it.

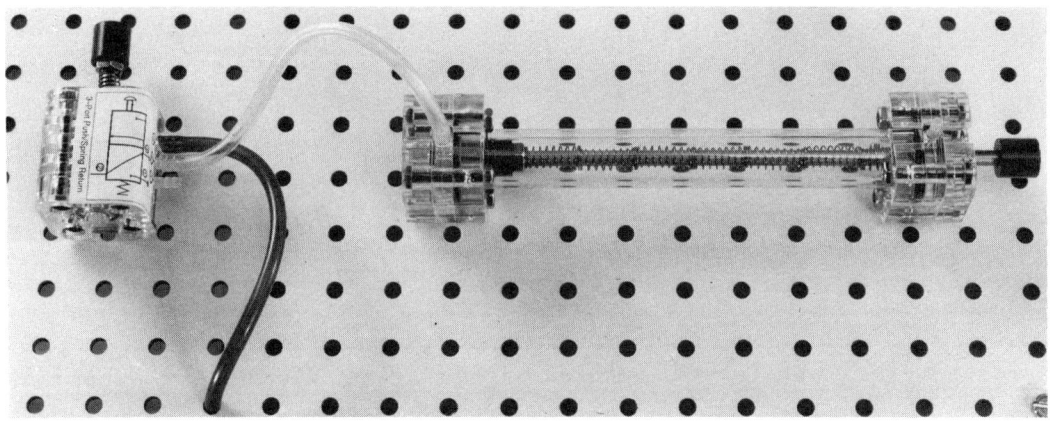

ASSIGNMENT 11.1

1) Copy and complete this truth table.

	Valve	Cylinder position
	OFF	Positive
	ON	

Connect another three-port valve **B** in the position shown below.

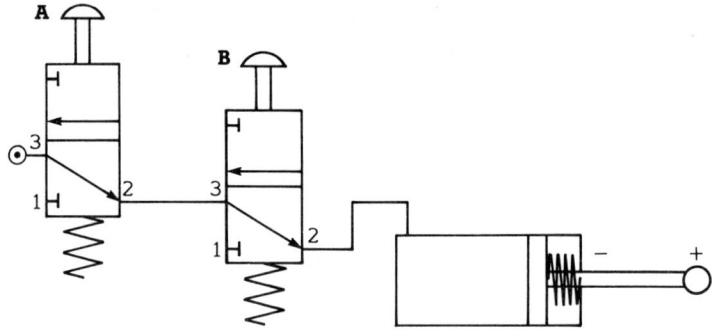

2) Copy and complete this truth table for the circuit above.

Valve A	Valve B	Cylinder position
OFF	OFF	Positive
ON	OFF	
OFF	ON	
ON	ON	

3) Add a third three-port valve **C** in the position shown below.

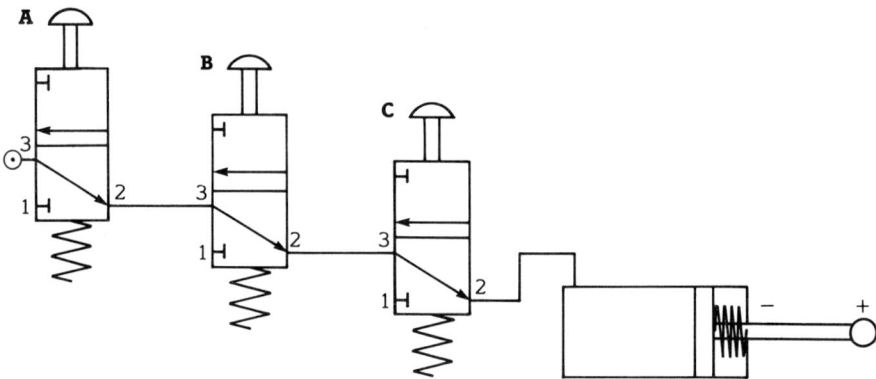

Operate the circuit again.

a) Has the new valve made a difference to the circuit's operation?

b) Describe an application for this circuit.

CHAPTER 12

Double-acting cylinders

Design problem

The circuit used to operate the sliding door in Chapter 9 has a major drawback. The cylinder can only go positive while a valve is operated. As soon as the valve is released the cylinder goes negative. Swift movement through the door is advised!

Design brief

Modify the circuit so that the door will remain open until a second valve is pressed.

Analysis/Research

A device is required that will open and close the door. The device must not move until operated by a valve. It must not return automatically.

A **double-acting cylinder** has a *power stroke* in *two* directions. Compressed air is used to move the piston both positive and negative. It can also be used to lock the piston firmly a the end of either stroke.

This is the symbol for a double-acting cylinder:

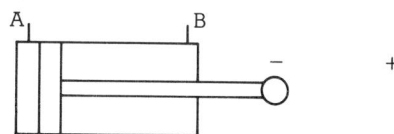

When air is put into connection A the piston extends to the positive position. Air in the cylinder ahead of the piston's movement can escape through port B. This air is called **exhaust air**.

To send the piston negative:

a) The air is disconnected from A.
b) The air is connected to B sending the piston back.
c) Exhaust air can escape through port A.

Design solution

Double-acting cylinders can be operated by two three-port valves.

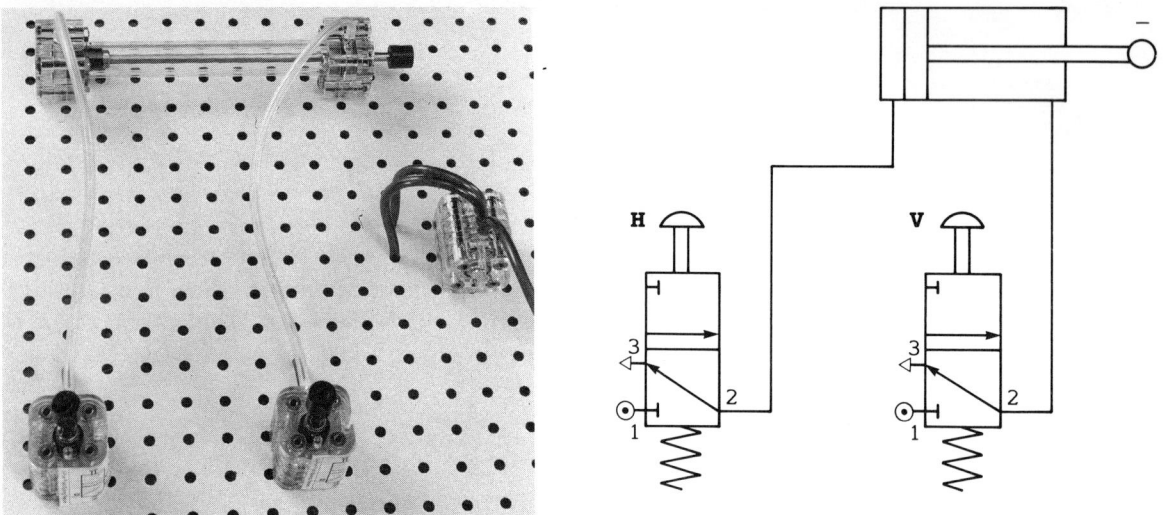

Evaluation

When valve **H** is operated the piston extends. When valve **V** is pressed the piston returns to the negative position.

OR gate controls need to be added to this circuit so that it can be operated from either side of the door.

ACTIVITY 12.1

Collect: 4 (push-button) three-port valves
            2 T connectors
            1 double-acting cylinder
            2 shuttle valves
            Connecting pipes

Draw the completed circuit that will allow the door to be operated from either side.

Assemble and test your circuit.

What happens if you now press valve **H** and valve **V** at the same time?

Can you explain why this happens? Work out an explanation with your teacher.

Cylinder speed control

Design problem

A double-acting cylinder operates very quickly. If it is being used to open and close a sliding door the speed at which it acts may damage the door mechanism or injure a person passing through it. The circuit drawn below can be used to operate a sliding door.

Design brief

Modify this circuit so that it will control the door's speed.

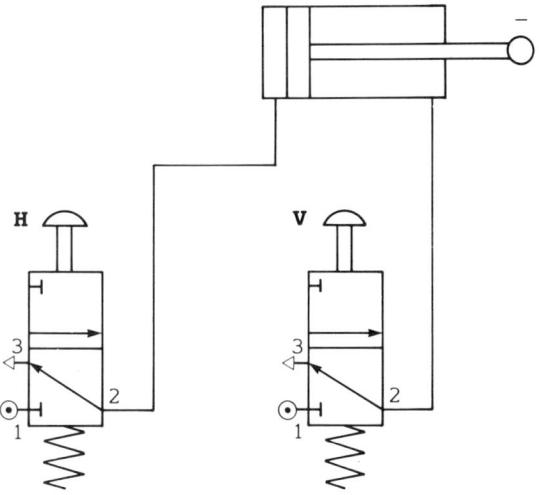

Analysis/Research

The speed of a piston's movement can be controlled using **flow regulators** or **throttle valves**. (Flow regulators can also be called **flow restrictors**.)

Flow regulators and throttle valves work by reducing the size of the hole that the air can flow through. They are adjustable so that the volume of air flow can be controlled. These valves are placed so that they control the exhaust air from a cylinder.

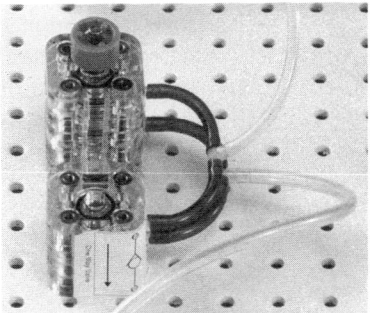

Flow regulator

A flow regulator restricts the flow of air in only one direction. The flow of air in the other direction is unaffected. This is the symbol for a flow regulator:

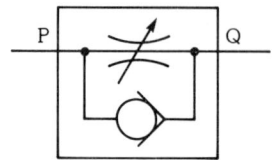

How flow regulators work

When the air flows into port P it has two possible paths.

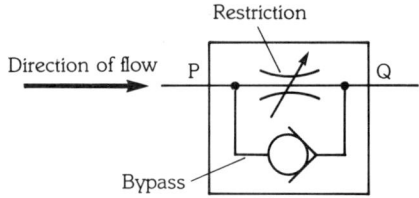

a) Some of the air takes the bypass route and forces the ball to seal the path (as in a shuttle valve):

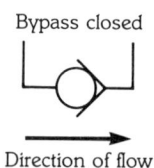

b) The rest of the air has to push its way through the restricted section (the upper path in the diagram). The amount of restriction depends on the setting of the control screw.

When the air flows into port Q it has two possible paths:

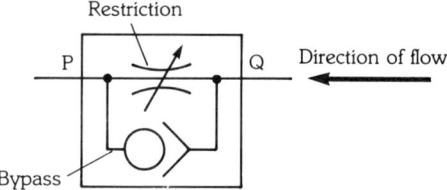

a) Most of the air follows the lower path, forces the ball to open this path and passes through the valve.
b) A very small amount of the air has to push its way through the restricted section (the upper path in the diagram), creating a pressure that acts back along the pipe.

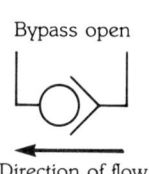

As the air from Q can flow through the unrestricted section the rate of flow is unchanged.

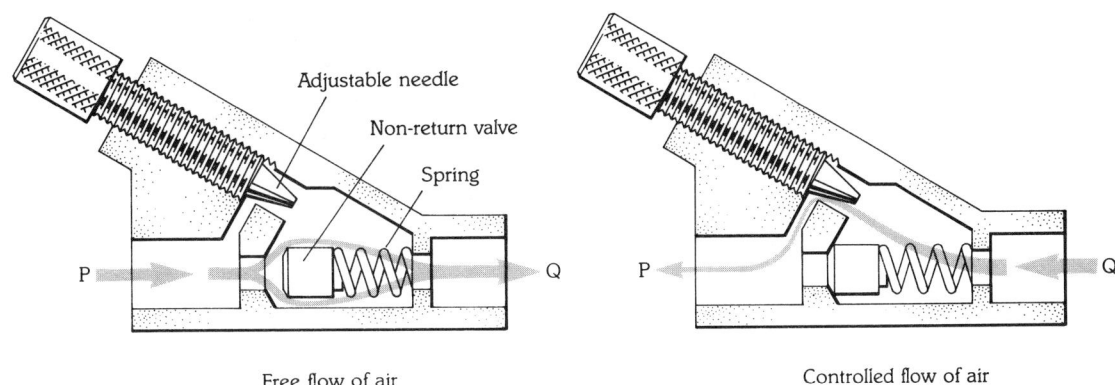

Adjustable needle
Non-return valve
Spring

P → → → Q

P ← ← ← Q

Free flow of air

Controlled flow of air

Placement of flow regulators

The correct placement of a flow regulator is shown below. Note that it will restrict exhaust air from the cylinder only on the positive stroke. (The arrow on the bypass route shows the direction in which restriction occurs.)

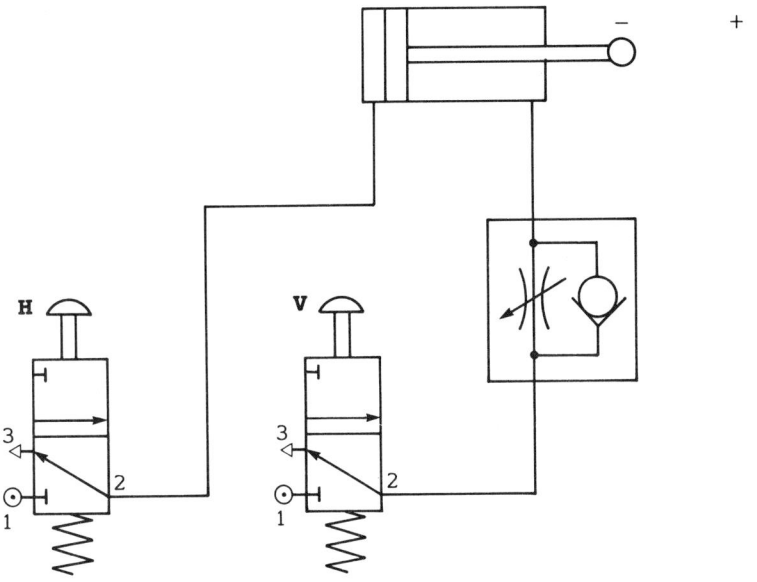

When valve **H** is pressed the compressed air flows through the pipe and sends the cylinder positive. The air that is trapped ahead of the moving piston has to escape through the flow regulator. Because of the way the regulator is connected into the circuit the air cannot escape quickly. This **back pressure** slows the speed of the piston.

The flow regulator can be adjusted to give different piston speeds.

To control the speed of movement in both directions an additional flow regulator will be needed.

Throttle valves

A throttle valve restricts the flow of air in both directions (from A and from S). This is the symbol for a throttle valve:

Air can flow through the valve in either direction. The controlled restriction works in both directions.

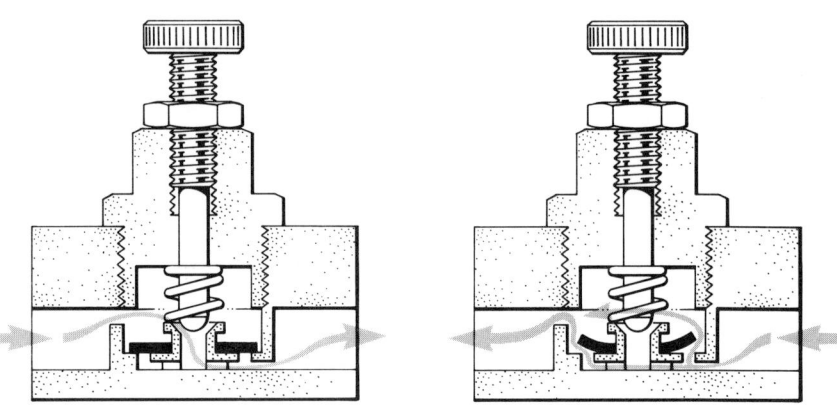

Design solution 1

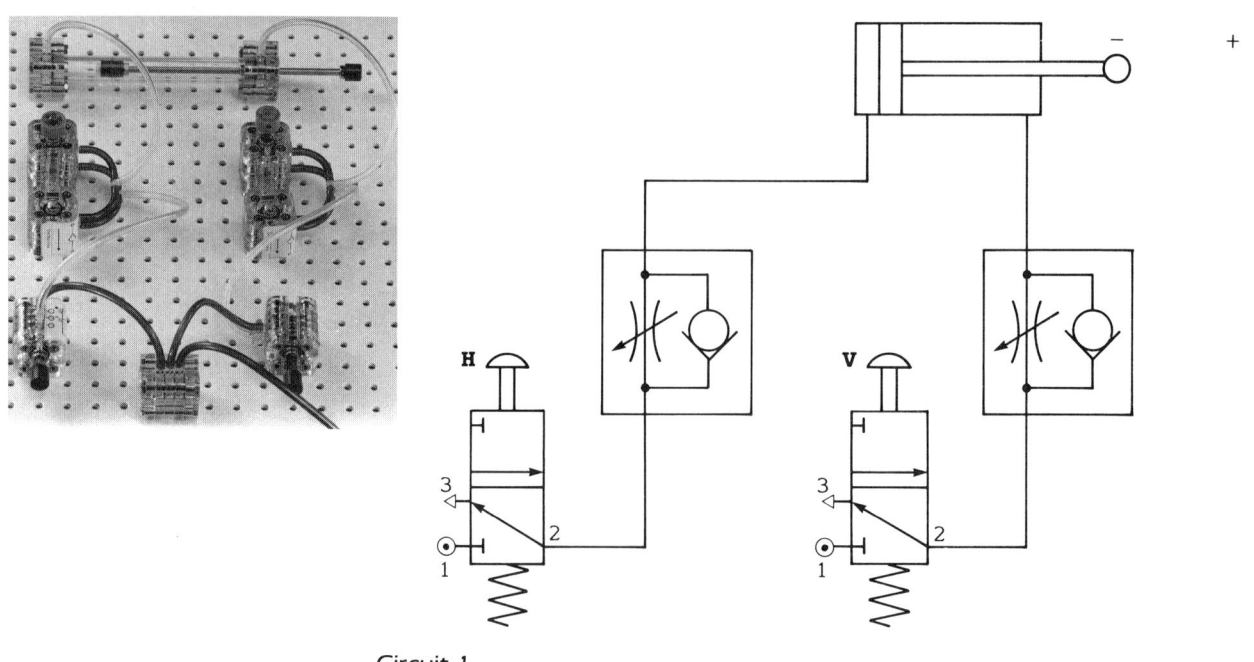

Circuit 1

Evaluation

When connected this circuit works correctly. The speed of piston movement is controlled by an appropriate flow regulator. Because the piston will stop as soon as a valve is released the three-port valves must be held **ON** for the complete stroke.

Design solution 2

This solution uses one throttle valve to control the piston speed in both directions:

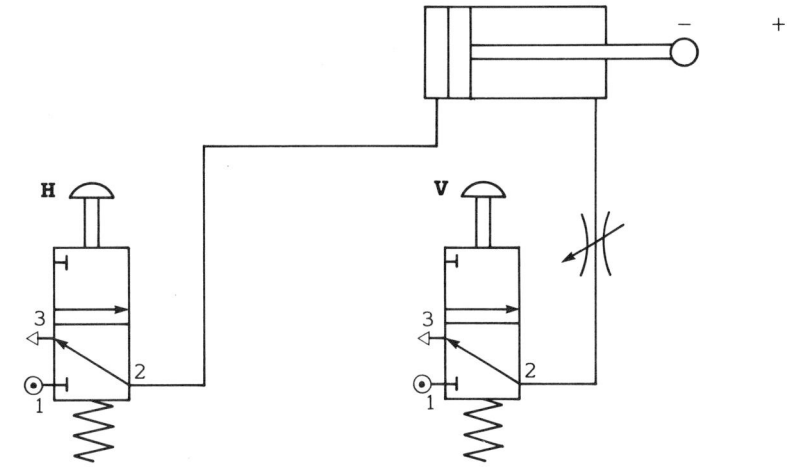

Circuit 2

Evaluation

This circuit works well when the piston is on the positive stroke. The movement is smooth and powerful. On the negative stroke the piston moves slowly and unevenly. This is because the throttle valve reduces the flow of air into the cylinder.

Design solution 3

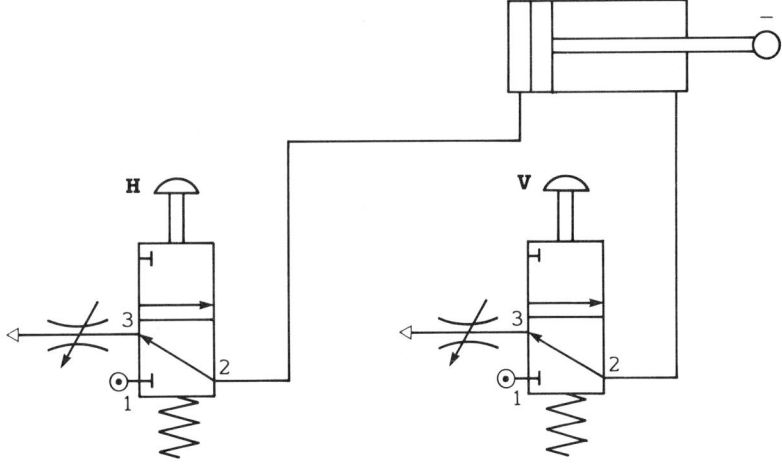

Circuit 3

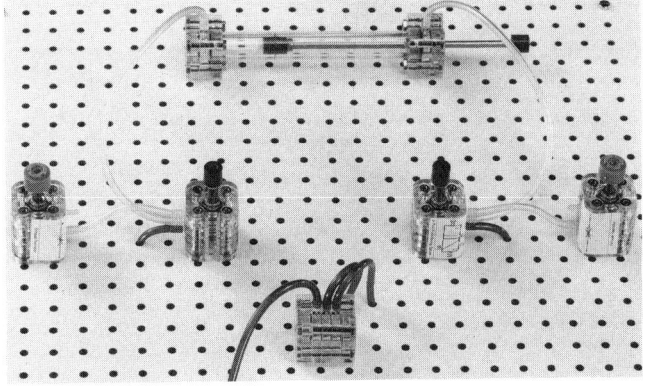

Evaluation

This circuit shows the correct placement of the throttle valves, on the exhaust air from the system. This circuit works with the required smooth and powerful action on the positive and negative strokes. Because the piston will stop as soon as a valve is released, the three-port valves must be held ON for the complete stroke.

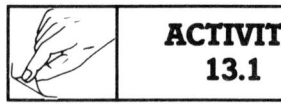

ACTIVITY 13.1

1) Assemble circuit 1.

   Adjust the flow regulators to make the piston move at a suitable speed for opening and closing a door.

2) Assemble circuit 3.

   Adjust the throttle valves to make the piston move at a suitable speed for opening and closing a door.

ASSIGNMENT 13.1

1) From a manufacturer's catalogue calculate the cost of circuit 1 and circuit 3.

2) Which circuit would you use to answer the design brief? Give three reasons for your choice.

3) What logic gate does this circuit resemble, OR, AND or NOT?

Five-port valves

In the last chapter we devised circuits to open and close a sliding door. One of these circuits is shown below.

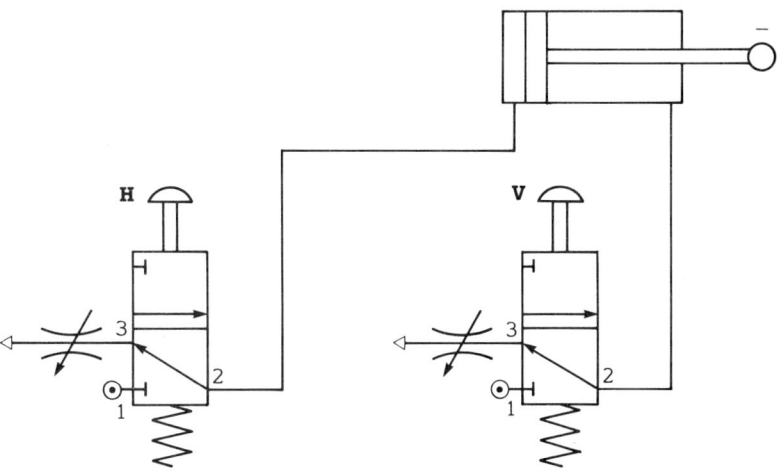

Design problem

This circuit has two main disadvantages:

a) The need to hold the operating valve ON from the start of the piston movement until the completion of the stroke

b) The large number of components which result in an expensive circuit.

Design brief

Find alternative components that will:

a) Improve the action of the circuit so that the valve does not have to be held ON all the time while the door fully opens

b) Reduce the cost of the circuit.

Analysis/Research

In the original circuit two three-port valves were used to control the action of the door. A **five-port valve** has the switching capability of two three-port valves operating together.

A five-port valve operated by a **lever set/reset switch** could be a suitable type of valve to select.

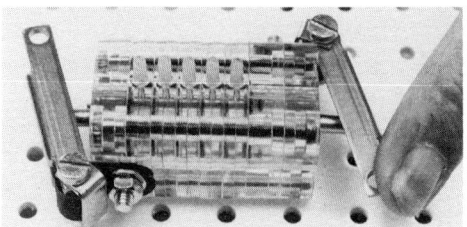

Lever set/reset five-port valve

A five-port valve has five main ports. They are laid out in this pattern:

4 (Compressed air out)
Channel A

2 (Compressed air out)
Channel B

5 (Exhaust air)
Channel A

1 (Compressed air in)

3 (Exhaust air)
Channel B

Five-port valves are not often used to turn OFF the air supply. They switch the compressed air output between ports 2 and 4, acting like OR gates. Their most common use is in the control of a double-acting cylinder.

Five-port valves have two states. We shall call these State A and State B.

State A

Compressed air is allowed to pass *in* through port 1 and *out* through port 2 to operate a piston:

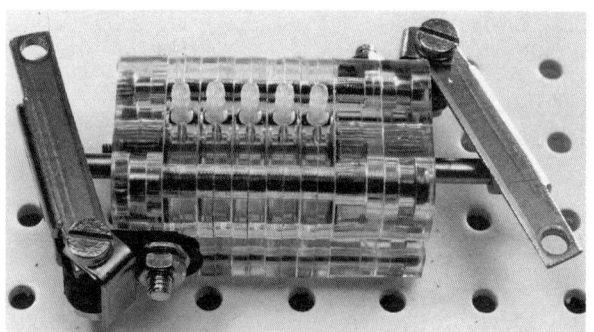 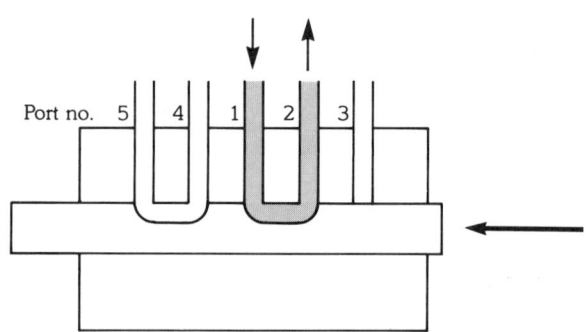

At the same time air can flow *in* through port 4 and *out* through port 5. This allows exhaust air from the cylinder to escape:

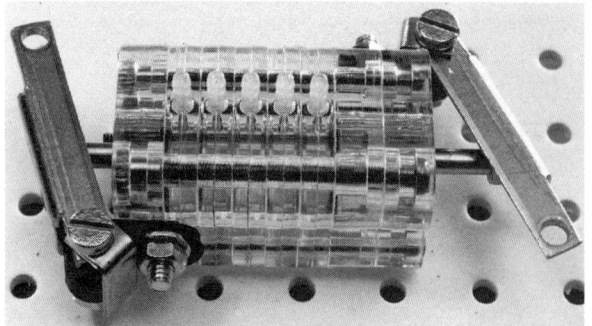 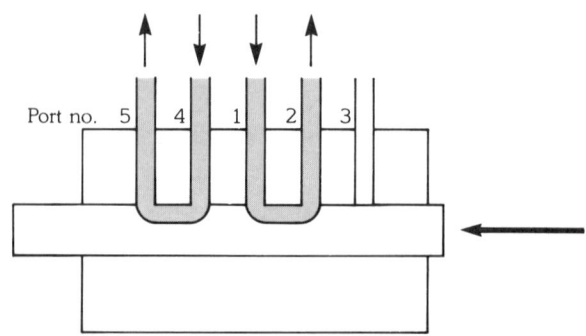

Note that port 3 is closed in this state.

When the valve has been changed by the lever set/reset switch the compressed air is allowed to pass *in* through port 1 and *out* through port 4 to operate a piston:

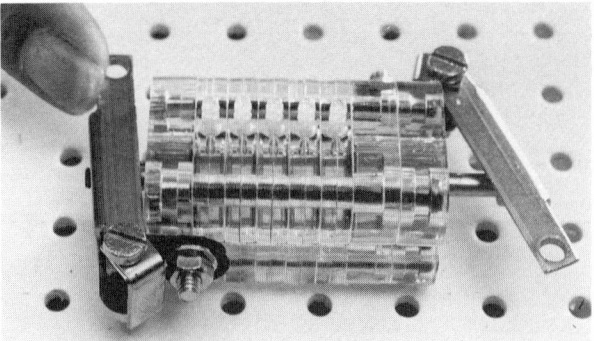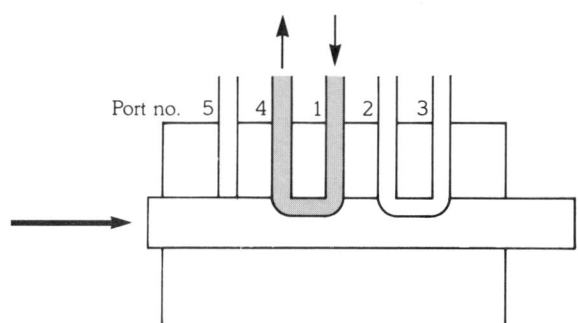

At the same time air can flow *in* through port 2 and *out* through port 3. This allows exhaust air from the cylinder to escape.

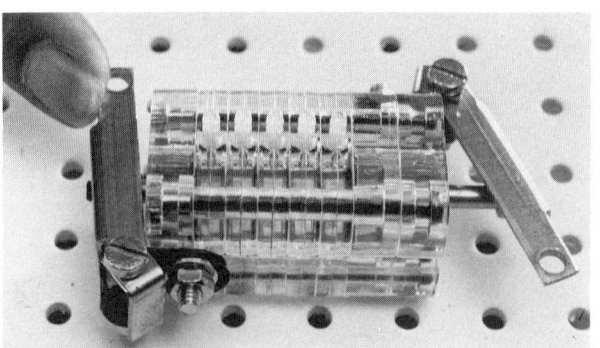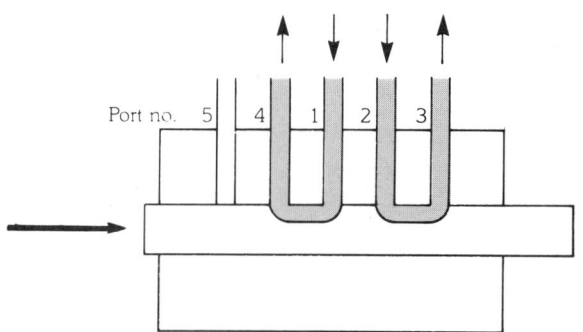

Note that port 5 is closed in this state.

Circuit diagrams for five-port valves

The circuit symbol for a five-port valve has to show several important features:

a) State A
b) State B
c) The type of switch used on the valve.

Like the three-port valve, this valve is drawn in two boxes showing the different air flow routes in each of its two states. The left-hand box shows state A and the right-hand box state B. The external connections are drawn on the sides of the box.

In state A ports 1 and 2 are connected. Ports 4 and 5 are also connected. Port 3 is closed.

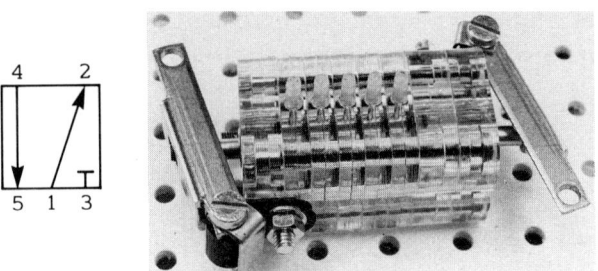

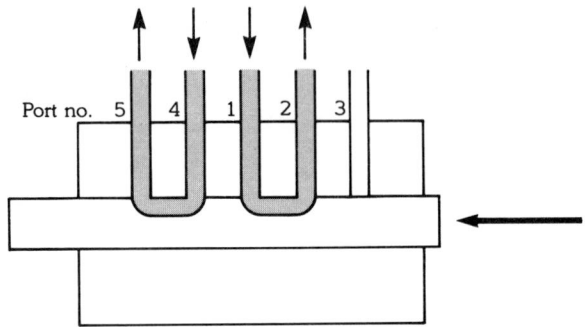

State A

The arrows show the direction of air flow between the connected ports.

In state B ports 1 and 4 are connected. Ports 2 and 3 are also connected. Port 5 is closed.

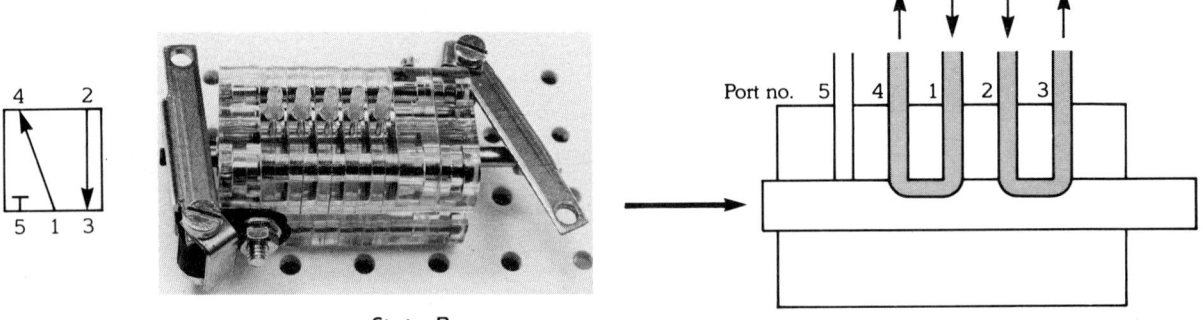

State B

Putting the state A and state B boxes together to form the complete valve symbol, we get:

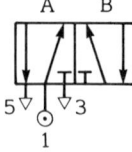

The symbols for the exhaust and the air supply are included.

The full circuit symbol for a lever set/reset five-port valve is drawn below:

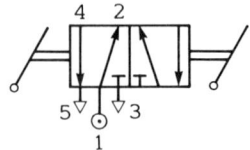

Note that it is normally drawn in state A.

ASSIGNMENT 14.1

1) Copy the diagrams of the five-port valves drawn below and explain in your own words what is happening.

2) Compare the drawings for the five-port valves below with drawings of the valves in Chapter 13. Note how the connections are drawn to match the valve's state.

Design solution

This valve can be connected to a double-acting cylinder to replace the two three-port valves used in the last design solution:

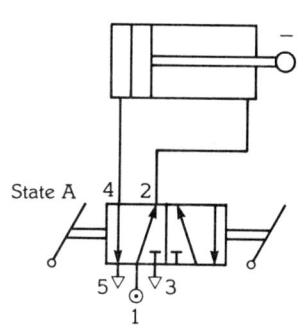

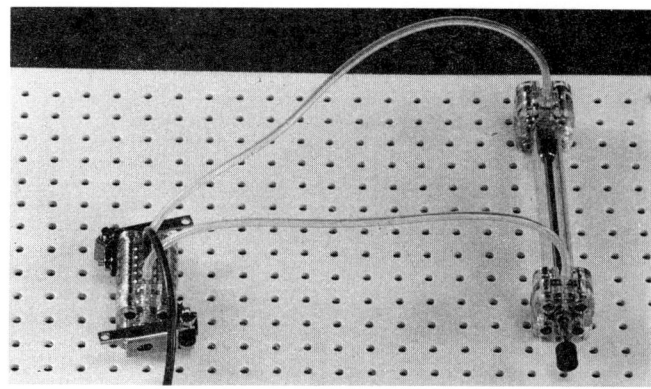

ACTIVITY 14.1

Collect the required components and assemble the circuit shown above.

1) Test to see whether it is easy to move the piston when it is either fully positive or fully negative. Explain your findings.

2) Add either throttle valves or flow regulators to the circuit to control the piston's speed of movement.

3) Draw a circuit diagram for your new circuit.

4) Write your own evaluation of your circuit in relation to the design brief.

Remote operation of five-port valves

Design problem

Technology rooms in schools have an electricity supply that can be instantly cut off to prevent accidents. A school technology room that does not have a similar cutoff for the compressed air supply may be putting pupils and staff at risk.

Design brief

Design a pneumatic emergency stop system for your school. The compressed air should remain disconnected from the room until a **reset valve** is operated. More than one working **stop button** should be provided for the room. There should be some kind of warning when a stop button has been pressed.

Analysis/Research

A pneumatic solution is appropriate because a compressed air supply already exists.

Additional stop switches are required.

A reset switch will be needed to reconnect the air supply.

A controlling valve will be needed for the compressed air.

The stop switch(es) should be suitable for operation by hand. The reset switch will need to be secure so that the problem can be sorted out before a responsible person resets the system.

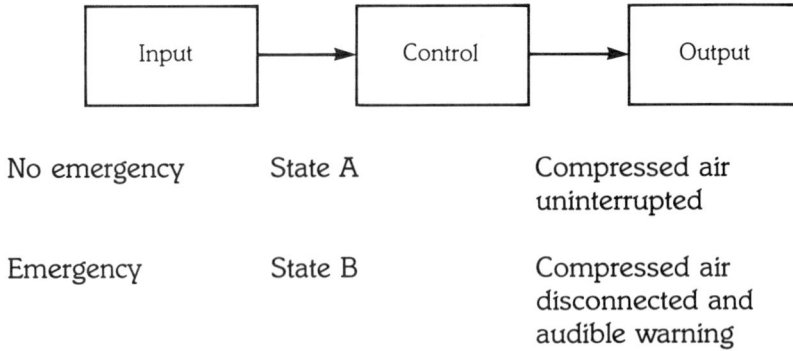

| No emergency | State A | Compressed air uninterrupted |
| Emergency | State B | Compressed air disconnected and audible warning |

From the evidence above we can see that the solution requires an OR gate. The most suitable valve for this solution is a five-port valve. The lever set/reset five-port valve is not appropriate because it must be operated directly and the stop and reset are combined in the same component.

A **double pressure-operated five-port valve** is a suitable component to use. It has five ports like the lever set/reset five-port valve, plus two signal ports. They are laid out in this pattern:

4 (Compressed air out) Channel A   **2** (Compressed air out) Channel B

Signal port **1(2)**   Signal port **1(4)**

5 (Exhaust air) Channel A   **1** (Compressed air in)   **3** (Exhaust air) Channel B

The **signal ports** are the connections used to control the operation of the five-port valve.

State A

Signal *in* 1(2)
High pressure *out* of port 2
Exhaust *out* of port 5

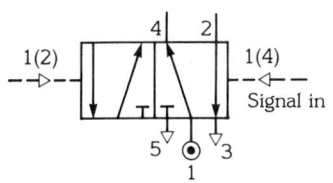

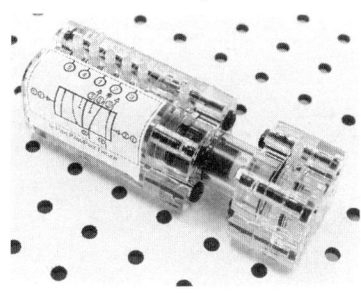

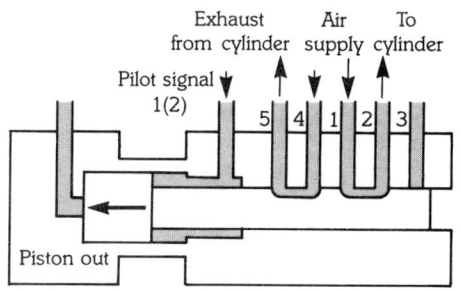

State B

Signal *in* 1(4)
High pressure *out* of port 4
Exhaust *out* of port 3

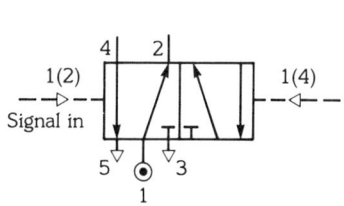

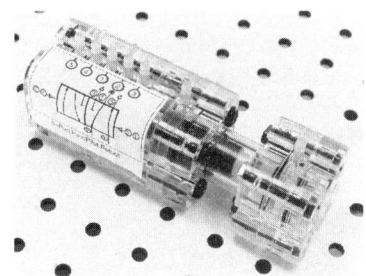

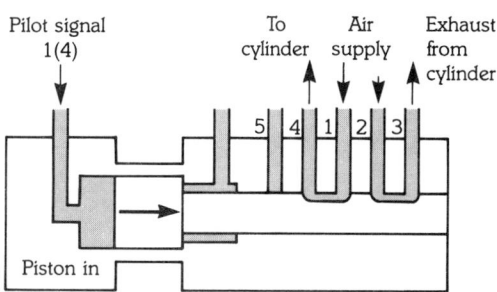

To change from one state to the other an air signal needs to be sent into the control ports 1(2) or 1(4). This can be delivered from a three-port valve. This is called a **pilot valve** when used in this way. Only a brief signal is required to change over the five-port valve's state.

Remember: Signal air *in* port 1(2) gives air *out* port 2.
                 Signal air *in* port 1(4) gives air *out* port 4.

Design solution (partly completed)

Note: Pilot signals are drawn as dashed lines: - - - - -

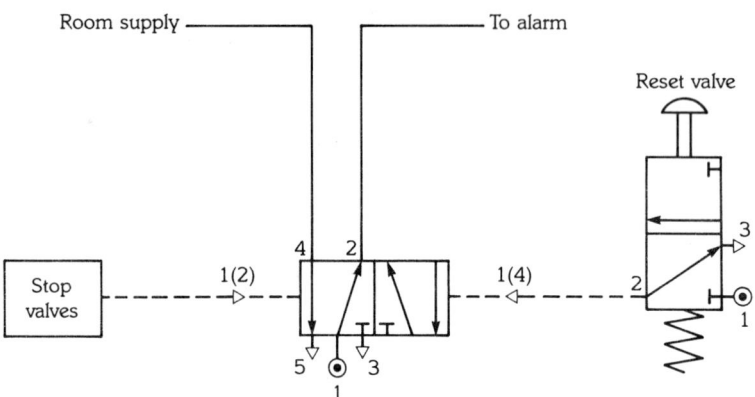

When the stop valve(s) have been operated the air supply is disconnected from the room and reconnected to an alarm, as shown above.

When the reset valve has been operated the air supply is connected to the room and disconnected from the alarm, as shown below.

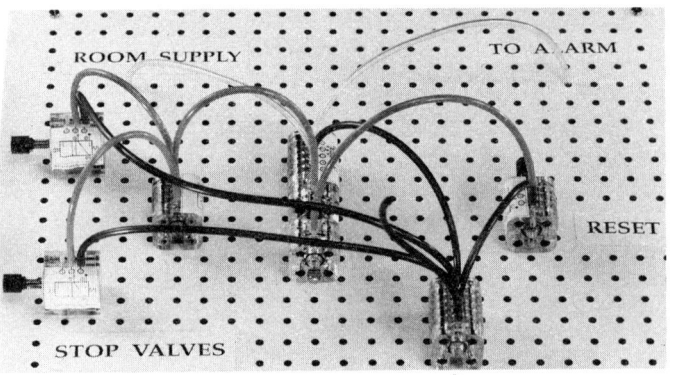

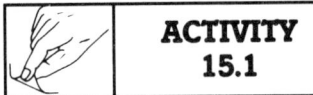

ACTIVITY 15.1

Using the circuit shown above as your starting point answer the following design brief.

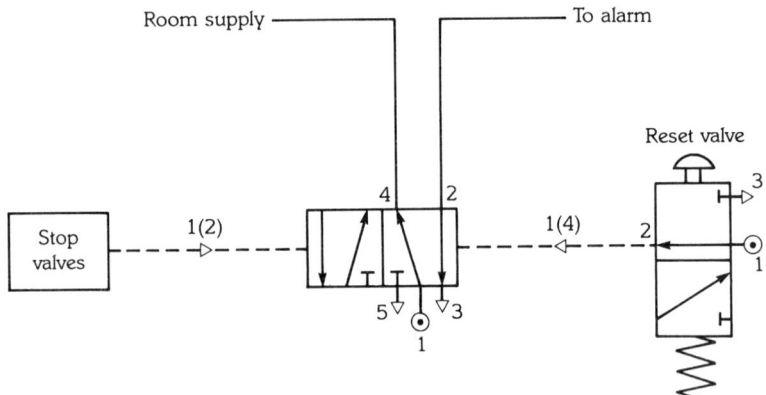

Design brief

Design an emergency stop system for a two-pupil pneumatic work station.

Safety point: The air supply to the alarm must not be left disconnected. All the main air supply will exhaust through this pipe when the system is triggered.

1) Write out a **specification** for your circuit (what you think it must do to completely solve the problem).

2) Draw alternative designs for your circuit that include pilot valves that are switched in different ways, for example foot operated or key operated.

3) Draw a circuit diagram of your solution. Add notes to explain your choice of components.

4) Describe the intended operation of your circuit.

5) Construct your circuit and test it.

6) Write an evaluation of your circuit comparing its performance with your specification.

Five-port valves operating double-acting cylinders

Design problem	The directors of a large company wish to check every vehicle that is being driven into their premises. To do this they intend to install a movable barrier at the entrance to their site.

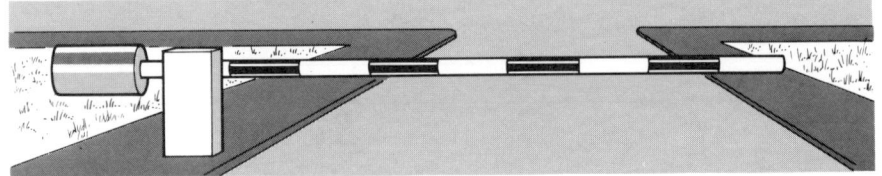

Design brief

Design a pneumatic circuit that will allow a security guard to open and close the barrier from some distance away.

Analysis/Research

A pneumatic solution is appropriate because the barrier is heavy and requires a large force to lift it.

A remote controlling system is required so that the barrier can be operated from a distance.

Operation of the barrier should be done by a double-acting cylinder because it needs to be controlled as it is raised and lowered.

The double-acting cylinder should be controlled by a five-port valve because it must be held for a time in the up position to allow a vehicle to pass. It needs to be lowered to stop unauthorised vehicles from passing.

Design solution 1

This circuit was used in Chapter 14. It could be suitable to use to raise and lower the barrier.

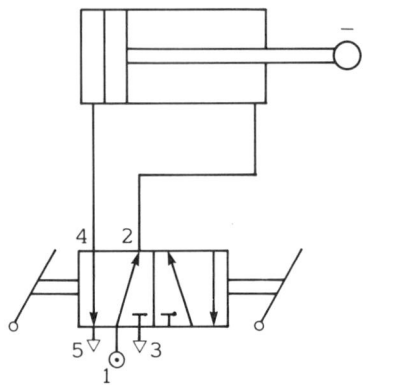

Evaluation

This circuit could be used to raise and lower the barrier if flow restrictors or throttle valves were used to control the speed of movement. The disadvantage of this design is that if the lever set/reset valve is not mounted near to the cylinder there could be large pressure losses in the connecting pipework. These would result in poor control of the barrier (it would not be possible to hold it firmly in position).

Design solution 2

If long connectors are unavoidable it is better to have them in the signal pipework whenever possible. This is because these lines do not carry air to components that are doing heavy work. We can modify the previous circuit to improve its performance by adding another five-port valve:

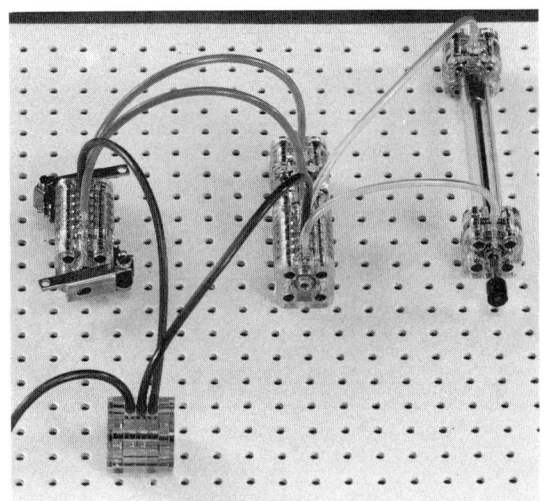
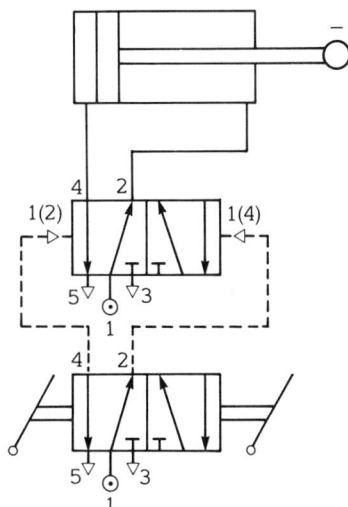

The new valve is a double pressure-operated five-port valve. It can be mounted close to the double-acting cylinder to ensure that the pressure losses are as low as possible. It is controlled from a remotely mounted lever set/reset five-port valve.

Evaluation

This circuit provides a remote controlling system operating the barrier from a distance. It also controls the raising and lowering of the barrier (it can be held in either extreme position).

ACTIVITY 16.1

1) Design and make a working model representing the barrier.

2) Modify the circuit for design solution 2 and then incorporate it into your model.

ASSIGNMENT 16.1

Read the evaluation of the last circuit again carefully.

Write an evaluation of your own circuit.

Automatic circuits

Design problem

In Chapter 16 we looked at a circuit to control a barrier. The barrier must be in the lowered position whenever a car arrives. The barrier is raised when the attendant wishes to allow the car to pass. The circuit could be improved by making part of its action automatic.

Design brief

Design a circuit that will automatically lower the barrier whenever a car approaches.

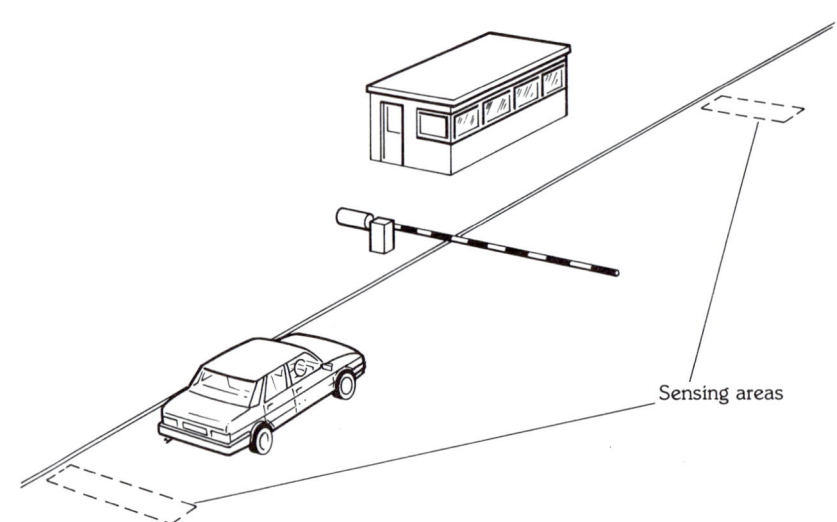

Sensing areas

Analysis/Research

The barrier needs to be operated by a double-acting cylinder controlled by a five-port valve.

A double pressure-operated five-port valve needs to be used so that the control valves can be located some distance away from the barrier.

Two **sensing areas** have been included in the road surface. Two sensors connected as an OR gate could trigger the lowering of the barrier. It would then be possible for the barrier to be:

a) Closed when the car arrives and also
b) Lowered when the car has passed through.

This would occur whichever direction the car was travelling in.

Design solution 1

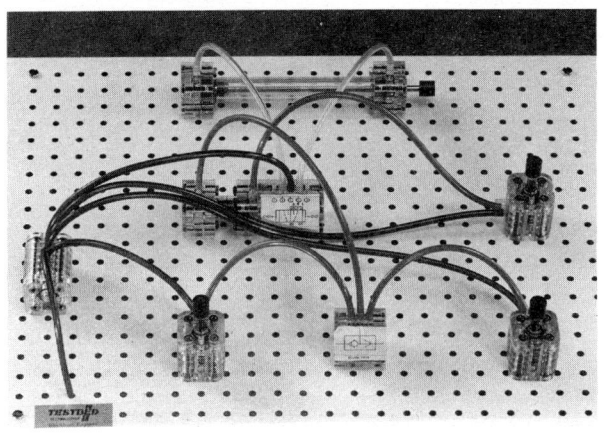

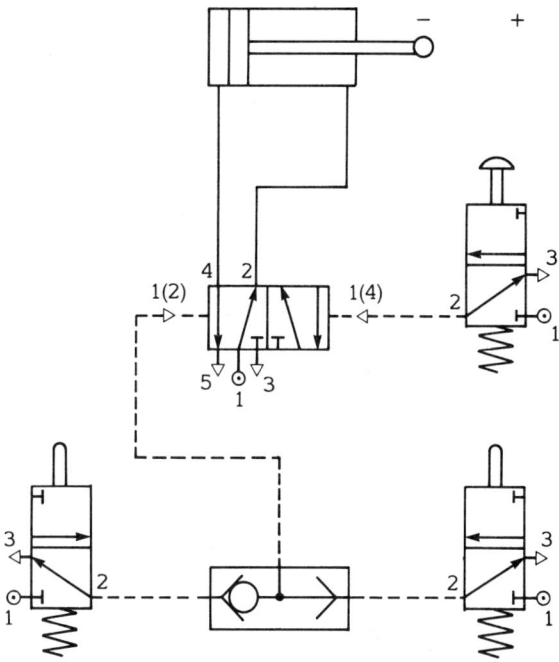

Evaluation

This circuit combines the OR gate used in Chapter 9 with the operation of a double-acting cylinder, controlled by a double pilot-operated five-port valve.

This solution will work satisfactorily when connected as a prototype, but installing it in practice will present many problems. For example:

a) The operation of the plunger valve in the sensing areas would require careful thought.
b) The installation of the piping would need to be done so that it did not intrude on the environment.
c) The whole system must be safe in use and must 'fail safe'.
d) The location of the circuit and the siting of the sensing areas are important to the correct operation of the system.

ASSIGNMENT 17.1

Design an addition to the circuit that will solve one of the following problems.

1) In the event of the plunger valves becoming dirty and jamming in the ON position the barrier will remain in the down position. Design an override device that will allow the attendant to raise the barrier.

2) If a car is approaching too quickly the attendant must be able to instantly reverse the action of the barrier to prevent an accident.
   Hint: See the emergency stop circuit, Chapter 15.

Draw the circuit for your solution.

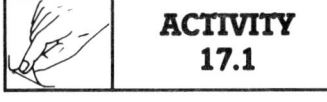
ACTIVITY 17.1

Connect the components for your circuit together. Test and evaluate it.

Semi-automatic circuits

Design problem

A small factory in an enterprise zone makes dowel-jointed shelving units. The workers want to improve their rate of production. The jointing holes in the units are drilled by an automatic system. A pneumatic cylinder controls the drill as it cuts the holes. The present pneumatic control circuit is shown below.

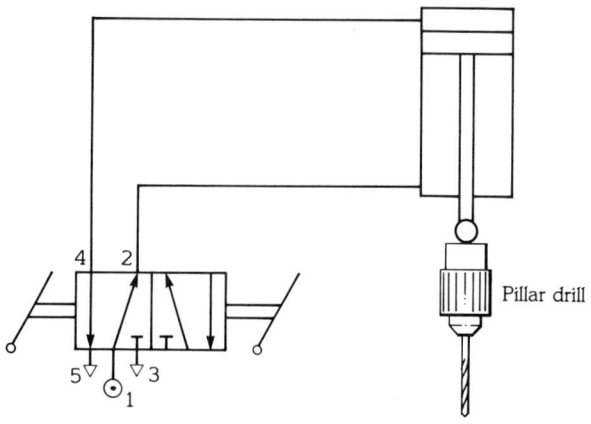

Pillar drill

Design brief

Design a circuit that will automatically lower the drill when the operator is ready to start drilling. The circuit should automatically raise the drill when it has cut to a preset depth.

Analysis/Research

The cylinder needs to be double acting and controlled by a five-port valve.

The system has to *detect* when:

a) The operator is ready to start drilling, and
b) The drill has reached the required depth.

The system must automatically *switch* the direction of the drill's travel when the hole has been cut to the correct depth.

The sequence of operations must be:

a) The operator places the board in position.
b) The start valve is pressed.
c) The drill is sent downwards to cut the hole.
d) The drill bit reaches a preset depth.
e) A signal is sent to the five-port valve.
f) The drill is sent back to the start position.
g) The operator removes the board.
h) The sequence is ready to repeat.

A double pressure-operated five-port valve can be switched by a foot- or hand-operated three-port valve to start the motion.

A valve is required to sense when the drill has reached the required depth of cut on its positive movement.

It must then send a signal that will change the state of the five-port valve. This signal is **positional feedback** – a feedback signal caused by a component moving to a certain position. This signal can be used to start the next operation.

The cylinder is then sent negative.

Position sensing

A plunger valve is placed so that it will be operated by a piston at the end of its positive stroke. A signal will be sent from that valve when the piston has reached the operating position.

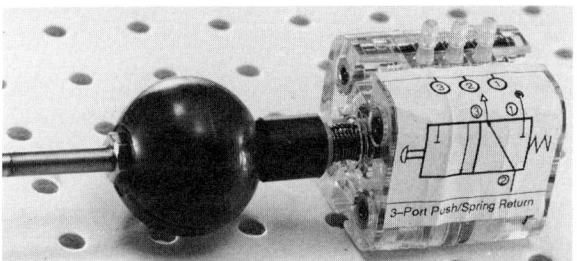

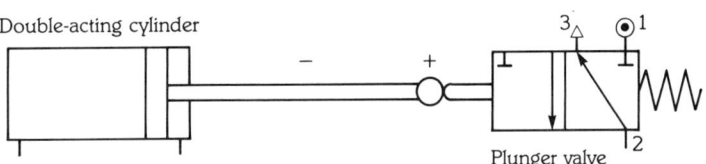

Double-acting cylinder

Plunger valve

A roller-trip valve can also be used to detect when the piston is in the positive position.

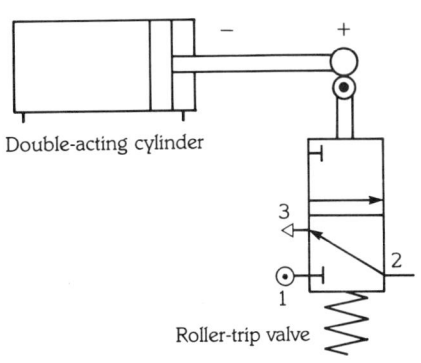

Double-acting cylinder

Roller-trip valve

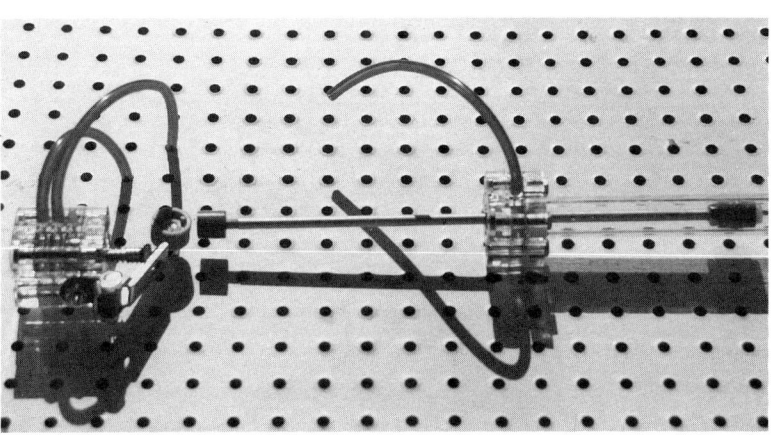

It can be used to detect the piston at the end of its negative stroke as well.

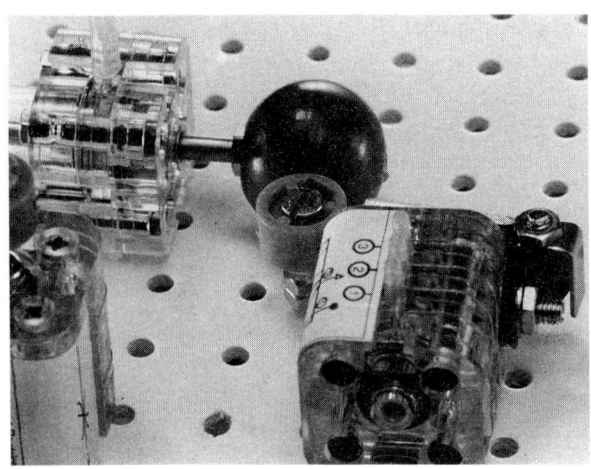

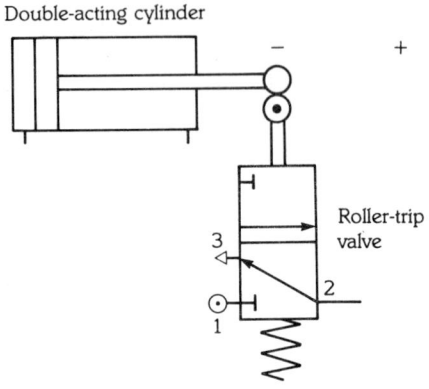

The choice of component will be made according to the available space and how and where it is to be fitted.

Design solution 1

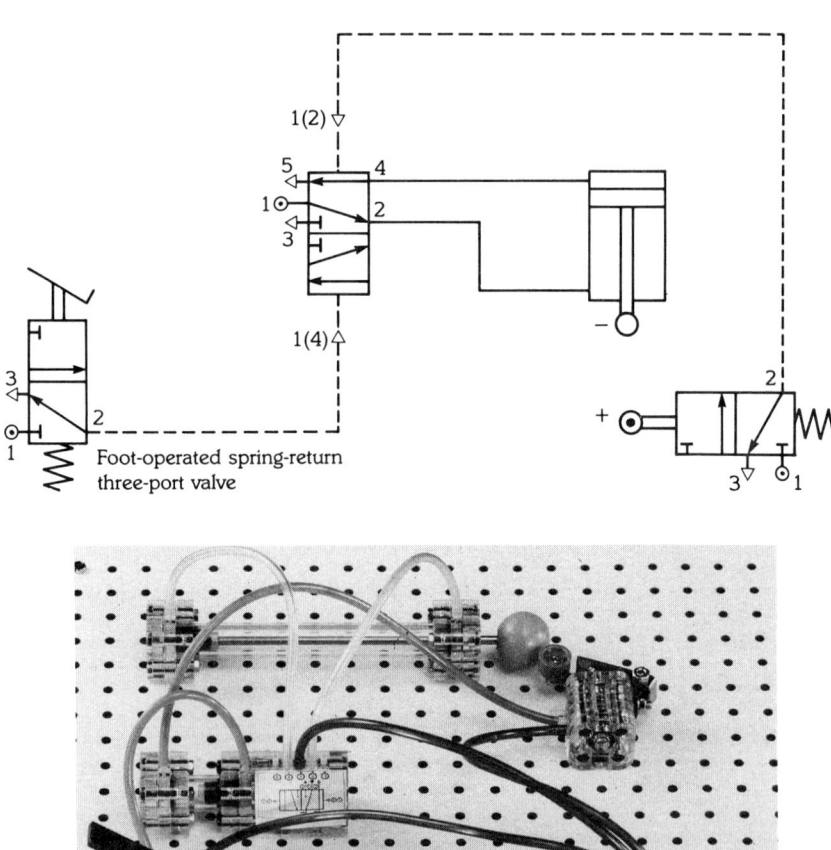

Design solution 2

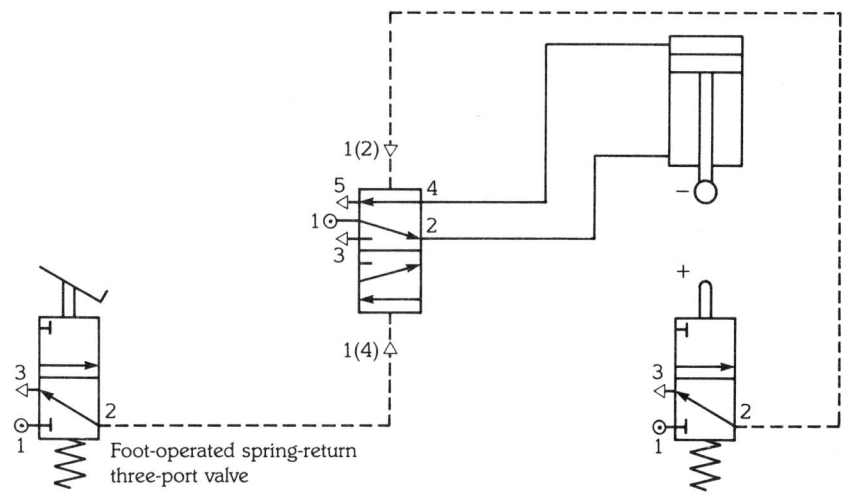

1(2)
5   4
1
3
2

1(4)

3
2
1
Foot-operated spring-return
three-port valve

+
3
2
1

**ACTIVITY
18.1**

Assemble both the circuits so that they can be seen operating side by side.

An alternative valve can be used in place of the foot-operated spring-return three-port valve.

Practical hints

You will probably find it necessary to **fine tune** the circuits to make them work properly.

Fine tuning means making small adjustments to a device or circuit so that it works as well as possible. It is a skilled job.

There are parts on some pneumatic valves and cylinders that can be moved to adjust the time and point at which switching occurs.

On some roller-trip valves there are two small screws.

In the roller-trip valve below, nut A can be adjusted to alter the angle of the arm, and nut B locks the backward movement of the arm.

B

A

On the cylinder the piston end can be moved on a screw thread to allow a small adjustment of the length of stroke.

Note: Some slotted base boards need to have the cylinders mounted at 90° to the slots. If cylinders are mounted parallel to the slots the circuits will suffer from **mechanical drift** due to the piston and the valve sliding along the slots. This will prevent the feedback signal being sent at the correct time.

ASSIGNMENT 18.1

1) Evaluate the two circuits you have made.

2) List any problems you think the circuits have.

3) Find ways of solving these problems.

4) Modify the circuits by including your suggestions.

5) Which final circuit solves the design problem best?

6) Design and draw a practical method returning the cylinder when the drill has reached the maximum depth of cut. The depth should be adjustable.

Design problem

The owners of a small fast-food shop want a machine that will rapidly slice small quantities of peeled potatoes into chips.

Design brief

Design an automatic system that will cut the chip shapes from peeled potatoes.

Analysis/Research

A pneumatic solution is appropriate because:

a) It is a clean system.
b) It is not as dangerous as electricity in combination with hot fat and moisture.
c) The pneumatic power can also be used for other tasks in the kitchen.

The potatoes are fed into position underneath the piston by a hopper system.

The potato is forced through the cutter by the action of the press.

Design solution 1

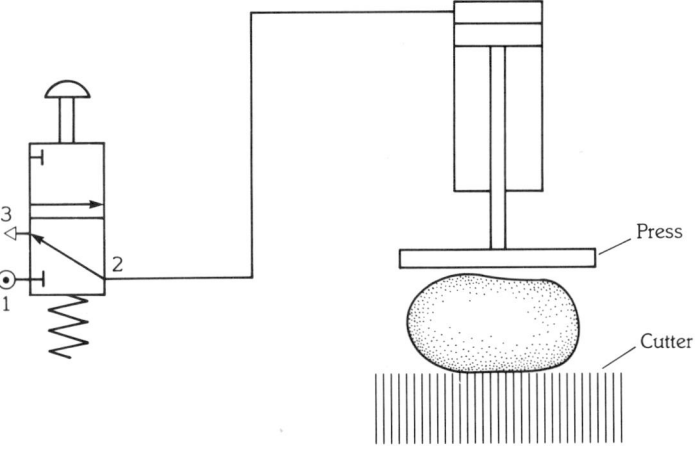

Evaluation

In this design the potatoes are cut when the three-port valve is operated. Someone is required to process all the potatoes manually.

There is a danger that mash will be produced rather than cleanly cut chips. Speed control cannot be achieved on the positive stroke because of the valve design.

This solution has a minimum number of components and would be relatively cheap.

This system cannot be adjusted very easily.

This system is not suitable for a fully automatic circuit. The design is difficult to modify. Further research is needed.

Analysis/Research

An automatic circuit is required that will repeatedly send the piston positive and negative.

As we saw in Chapter 18 it is possible to have an automatic negative stroke (instroke) using a three-port valve to sense when the piston has reached the end of its travel.

This circuit sends the piston positive when the push-button spring-return three-port valve is pressed. The plunger-operated spring-return three-port valve sends the piston on its instroke.

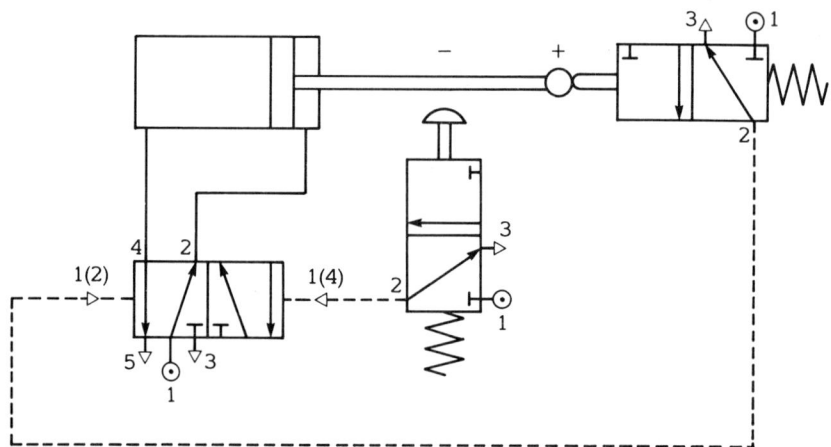

ASSIGNMENT 19.1

Describe in detail the operation of this circuit. List the sequence of operations as follows and complete the missing sections:

1) Push-button spring-return three-port valve ON.

2) Double pilot-operated five-port valve connects port 1 to port 4.

3) Piston is sent positive.

4) ?

5) ?

6) ?

7) Repeat sequence.

By replacing the push-button spring-return three-port valve with a roller-trip spring-return three-port valve and positioning it so that it is activated by the piston when it is fully negative the circuit can be made fully automatic.

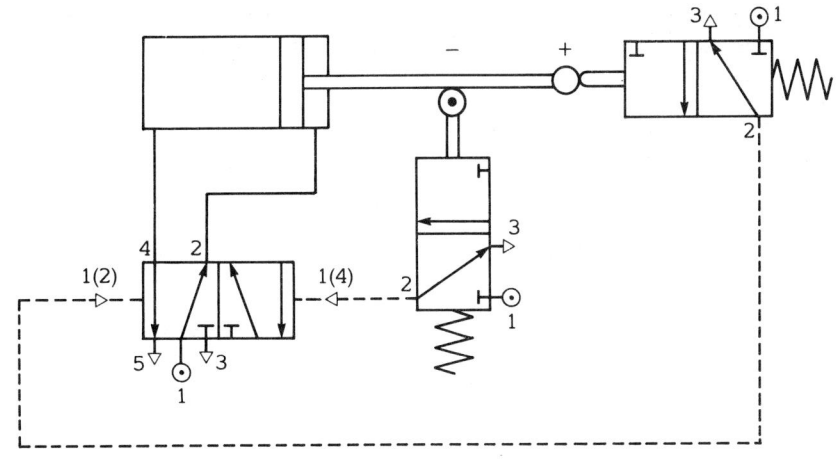

A suggested layout is given below.

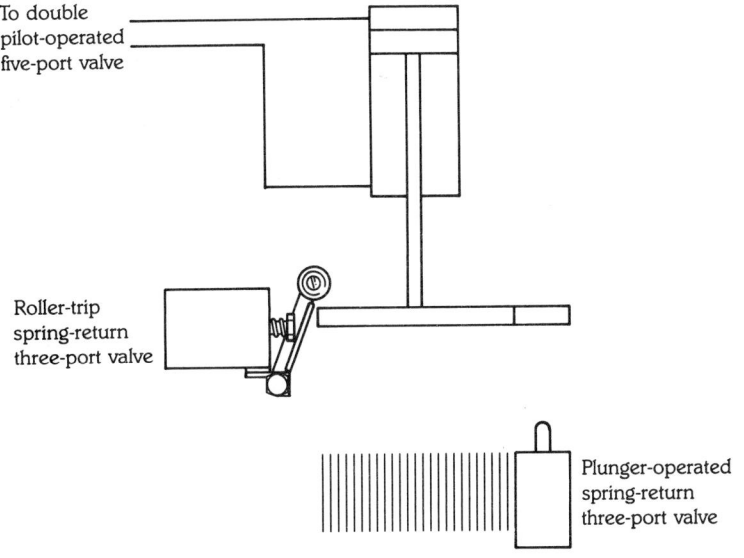

To double
pilot-operated
five-port valve

Roller-trip
spring-return
three-port valve

Plunger-operated
spring-return
three-port valve

 **ACTIVITY 19.1** Construct this circuit. Evaluate its performance and make modifications to allow the potatoes to be cut correctly.

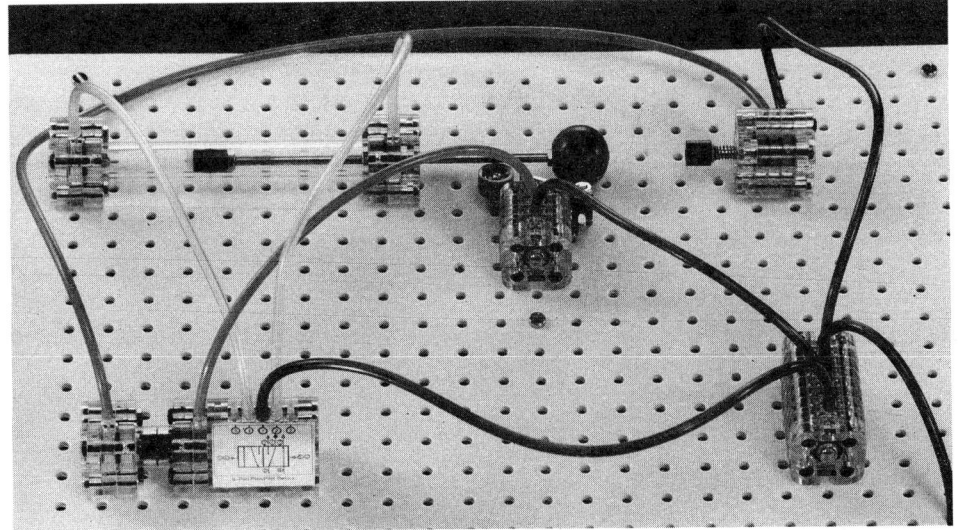

The circuit used in Chapter 19 could have been modified with a flow restrictor or throttle valve to control the speed of the positive stroke (outstroke).

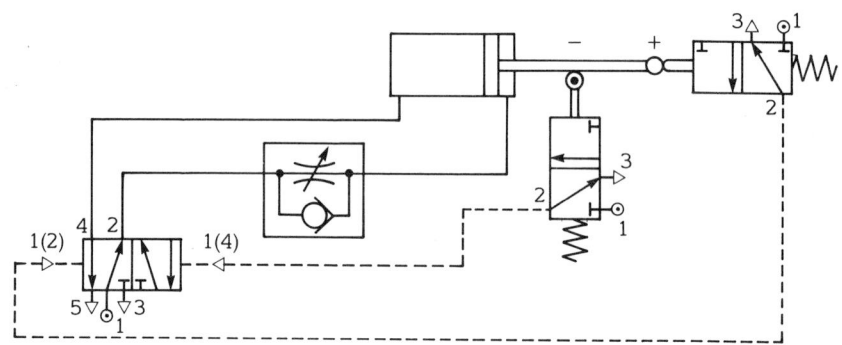

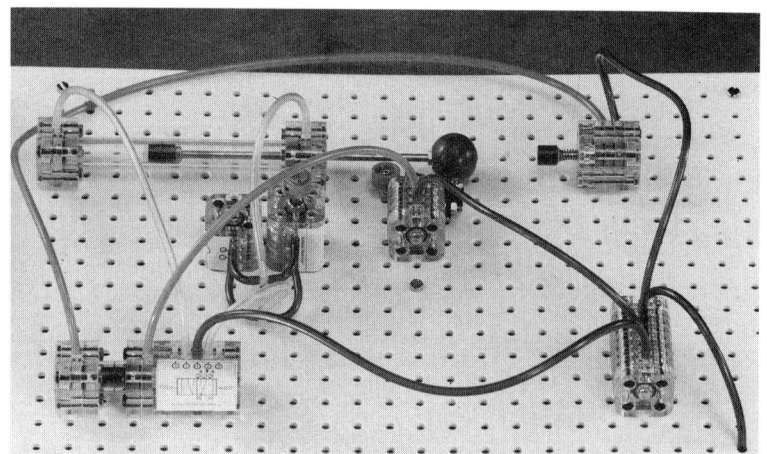

| **Design problem** | This circuit cannot sense when a potato is present. It relies on a regular flow of potatoes to run efficiently. |

Design brief   Modify the circuit above so that it can sense the presence of a potato.

Analysis/Research   A weight- or size-sensing system would be complicated.

The valves will become covered in potato residue and will need cleaning to prevent malfunction.

Sensing the presence of an object can be achieved by obstructing a jet of air. This system is called an **air bleed**.

An air bleed is a piece of tubing without any connections on its open end. Low-pressure air is supplied via a flow regulator. The air is allowed to escape from the pipe. If the air leaving the tube is obstructed in some way a back pressure builds up. This increased pressure can be detected by a **pressure-sensitive spring-return three-port valve**. This three-port valve can then be used to operate another pneumatic device using the high-pressure signal from port 2.

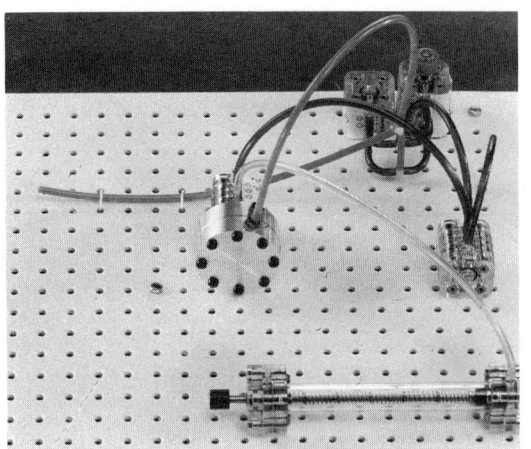

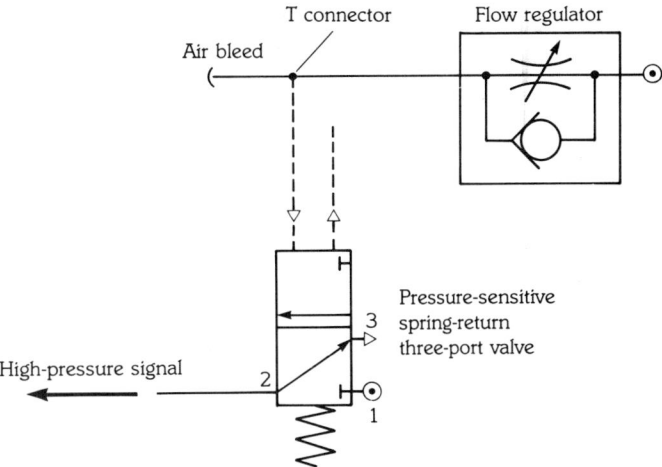

The pressure-sensitive spring-return three-port valve has two inputs that are controlled by changes in low pressure:

a) One that detects an increase, and
b) One that detects a decrease (a vacuum).
When a change is detected the valve switches states.

Construct this circuit.

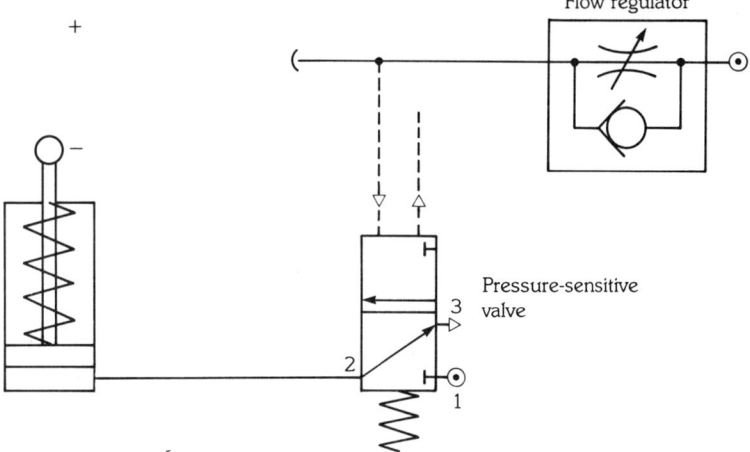

Adjust the flow regulator so that the pressure-sensitive spring-return three-port valve is on the verge of switching. This fine tunes the air bleed circuit so that it is operating in its most sensitive state.

Experiment with different low pressures and observe how they affect the performance of the circuit.

Note: A single-acting cylinder has been connected to the high pressure output from the pressure-sensitive spring-return three-port valve for safety reasons.

An air bleed can be used to detect the presence of a potato. It can replace the plunger-operated spring-return three-port valve in the design for the chip cutter.

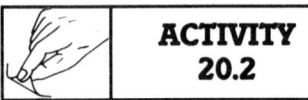

ACTIVITY 20.2

1) Draw the correct circuit diagram for the modified design.

2) Assemble the new circuit.

3) Fine tune it so that it works correctly.

4) Add another air bleed sensing system to detect the completion of the piston instroke.

5) Evaluate this circuit. Write out your findings.

In Chapter 9 we examined a solution to this design brief:

'Design a pneumatic control system that will allow the coldroom door to be opened from either side.'

The solution led us to concentrate on an OR gate valve arrangement. But the control of the door movement was far from ideal.

Design problem

The sliding entrance door in a supermarket is opened pneumatically by the operation of a three-port valve. When the door is fully open it trips a second three-port valve which automatically shuts it.

Despite slowing down the speed of the door's movement, the manager of the shop thinks that the cycle of opening and closing is too short. It does not give customers very much time to pass through the door before it shuts on them.

The existing pneumatic control circuit is drawn below.

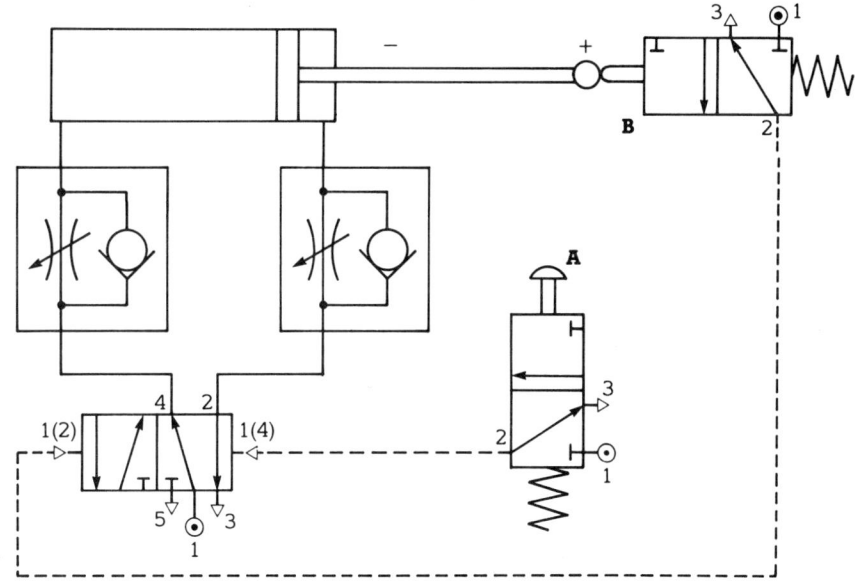

ASSIGNMENT 21.1

Write an analysis of the operation of the existing door control circuit. List the sequence of operations.

Design brief

Design modifications to the circuit that will leave the door open for a longer time so that the customers have more time to pass through.

Analysis/Research

The easiest method of slowing down the operation of a circuit is to use piston speed control. This is already present in the circuit.

When valve **B** is pressed it sends an air signal to port 1(2) of the five-port valve. This valve switches over immediately sending the piston negative. A delay is required in the operation of the circuit before the five-port valve is switched over. This is done by adding a **time delay** system to the circuit.

A five-port valve needs a reasonable pressure of air to switch it from one state to the other. If the signal air flowing to the 1(2) port is restricted then a short delay will occur before there is enough pressure to switch the five-port valve over.

We have already seen how a flow restrictor can be used to control the rate of air flow in a circuit. This would be a suitable component to use, inserted in the signal airline like this:

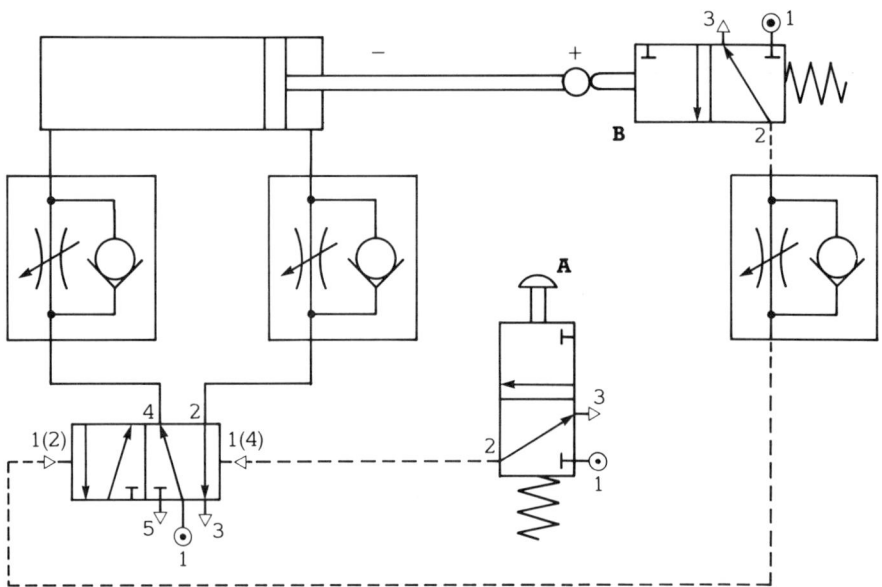

Evaluation

The adjustment of the flow restrictor in the signal line is critical. To achieve any delay at all it must be almost fully closed.

This modification to the circuit does produce a small time delay before the piston goes negative. A longer time delay is needed. This is achieved by adding a **reservoir**. It is placed between the flow regulator and the five-port valve on the signal airline.

Reservoirs

This is the symbol for a reservoir: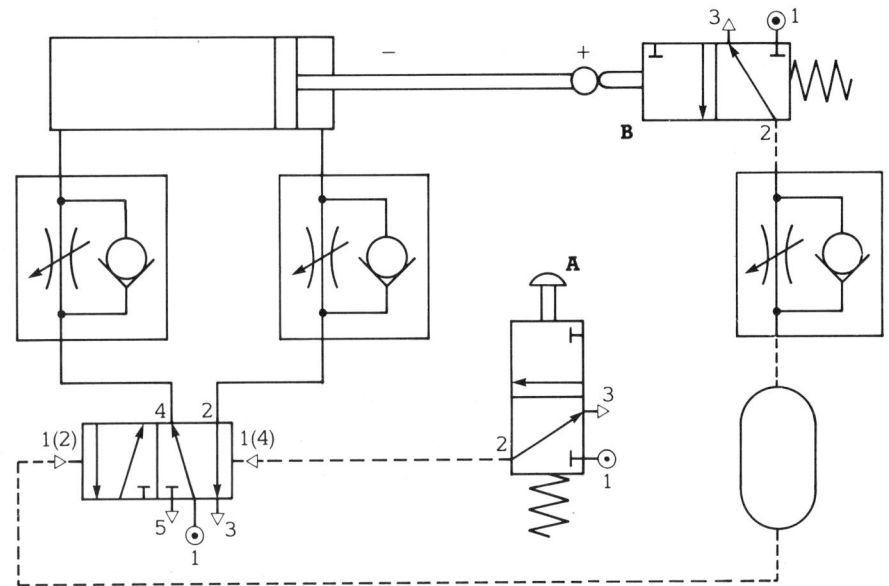

A reservoir is a device which needs a large volume of air pumped into it to achieve the pressure required to operate the system.

Adding a reservoir, the circuit now looks like this:

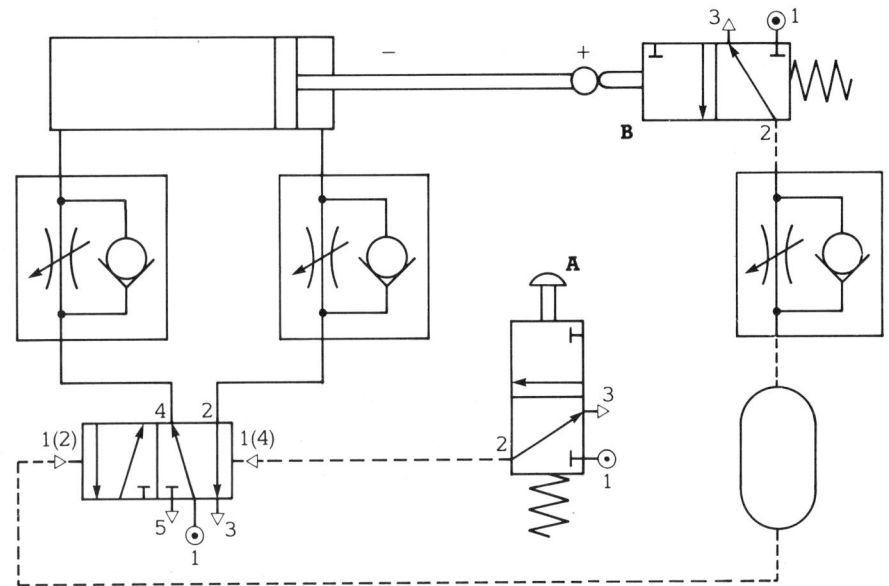

When there is enough pressure in the reservoir, the five-port valve is tripped and the piston goes negative. The time taken for the pressure to build up in the reservoir depends on the amount of restriction applied to the flow regulator.

> A large restriction = a large time delay
> A small restriction = a small time delay

Another factor that can change the time it takes to pass a signal to a five-port valve is the **reservoir volume**.

If a reservoir of twice the volume is inserted it will take approximately twice as long for the pressure to build up and operate the five-port valve, if the flow regulator is left unchanged (see overleaf).

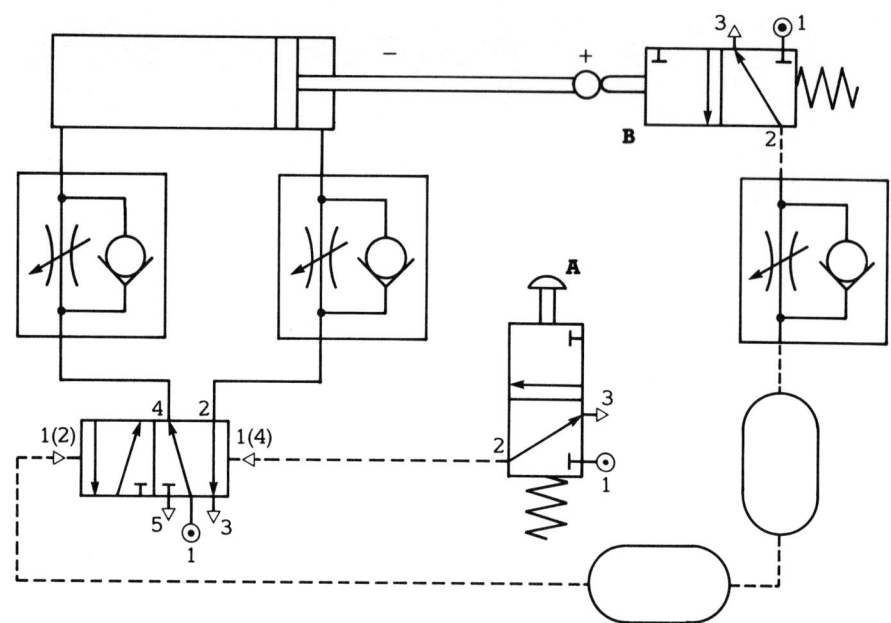

A guide for the amount of time delay in a circuit is:

A large restriction + a large reservoir volume   =   a long time delay
A small restriction + a small reservoir volume   =   a short time delay

ACTIVITY
21.1

1) Assemble the circuit shown above.
2) Evaluate its performance in terms of the design brief.

Design problem

In order to remain competitive a small company which produces ready-to-assemble furniture needs to speed up the rate of production of certain basic wooden parts. It also wishes to experiment with the use of pneumatic power as an alternative to electrically driven equipment.

A basic part is shown here.

Design brief

Design and make a model of a pneumatic system that will allow the part drawn to be positioned, drilled and released in one sequence of operations. The device must be simple to maintain and cheap to install if it is to be approved by the chairperson of the company.

Analysis/Research

A compressor is already available in the workshop to clean existing machinery.

The sequence of operations for the machine must be:

a) The part is pushed into position and held by guides.
b) The drill is lowered to cut the hole.
c) The feed system allows the next piece of wood to drop into position.
d) The drill is raised.
e) The next piece of wood is pushed into the drilling position, removing the machined piece.

In order to achieve stages (a) to (e) the movement of both the drill and the feed system must be regulated by pneumatic cylinders.

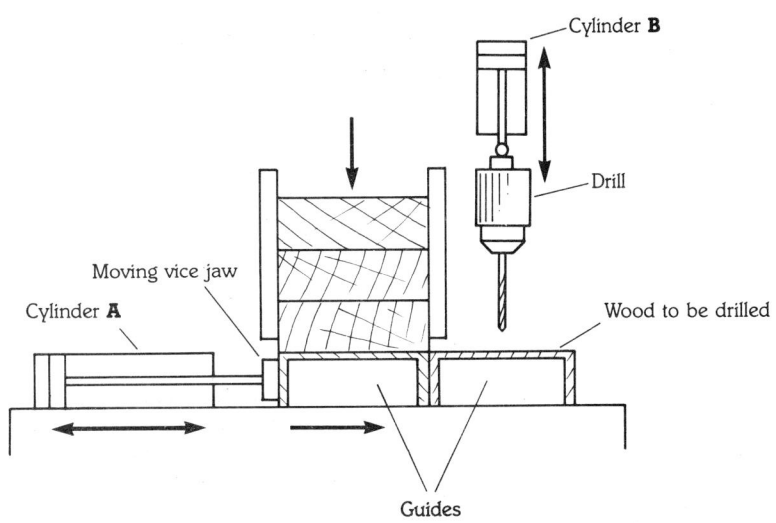

Stage 1
Cylinder **A** goes positive and pushes the wood into the drilling position.

Stage 2
Drilling cylinder **B** goes positive and drills the hole.

Stage 3
Cylinder **A** goes negative, allowing the next piece of wood to drop in front of the feed cylinder.

Stage 4
Drilling cylinder **B** goes negative, withdrawing the drill from the work.

In pneumatic notation the sequence of operations can be listed as:

> START
> A+
> B+
> A−
> B−
> STOP

To achieve this sequence of operations two cylinders, feed cylinder **A** and 'drilling' cylinder **B** are each connected to a double pilot-operated five-port valve in the normal way.

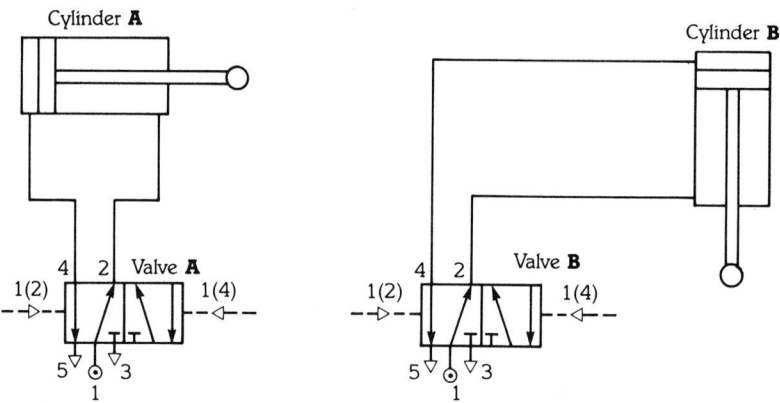

The part is placed in position. The sequence is then started by pressing a push-button three-port valve (**Z**) which is connected to five-port valve **A** at port 1(4).

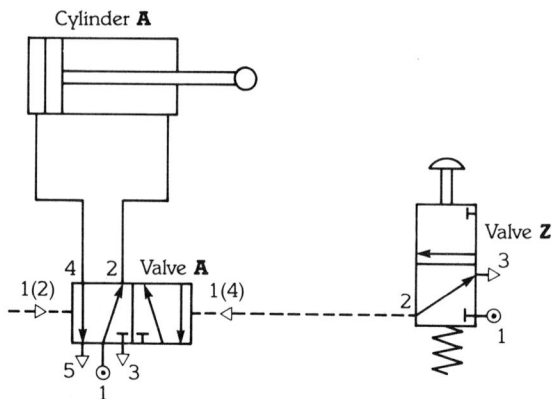

When cylinder **A** goes fully positive it not only feeds the wood into place but also operates a plunger valve (**a+**). This plunger valve sends a signal to port 1(4) of valve **B**.

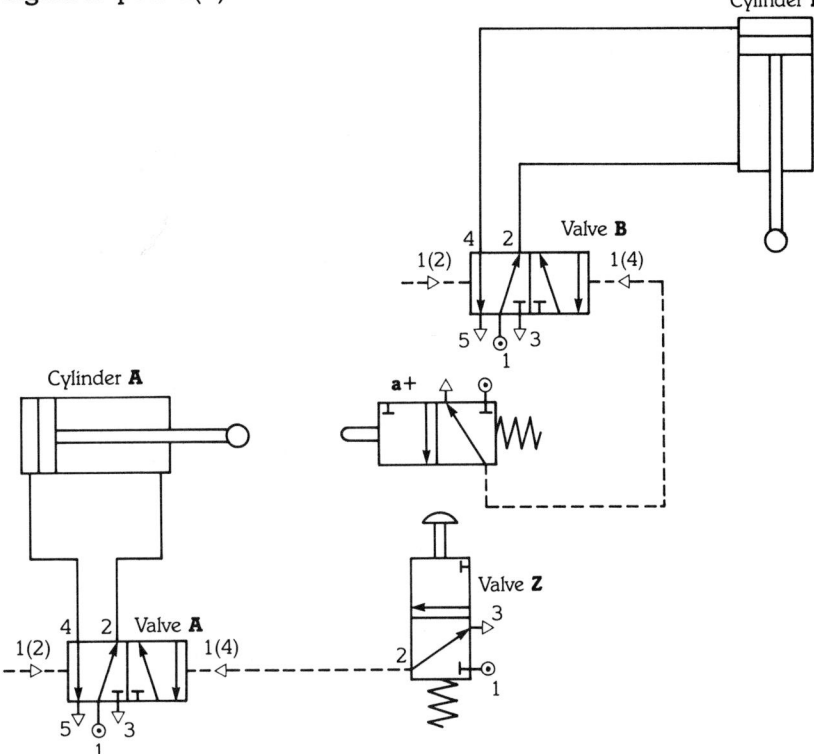

Drilling cylinder **B** is then sent positive.

At the end of the positive stroke drilling cylinder **B** operates a plunger valve **b+** which is connected to port 1(2) of valve **A**. This causes valve **A** to switch states and sends feed cylinder **A** negative.

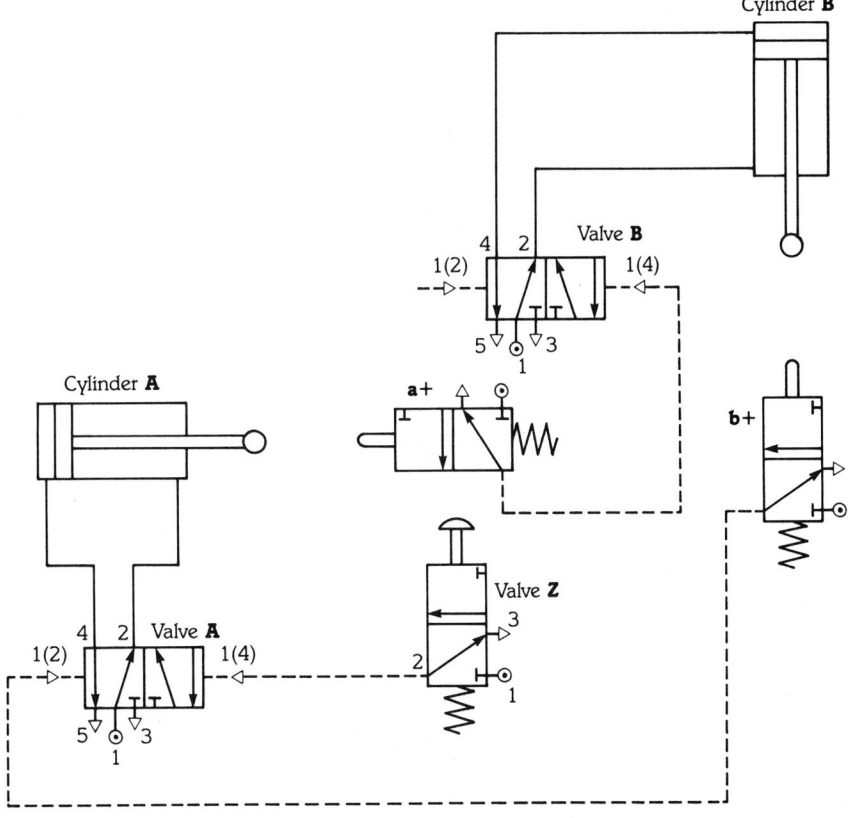

At the end of its instroke piston **A** operates a roller-trip valve **a−**. This valve is connected to port 1(2) of valve **B**. The signal from **a−** causes valve **B** to change state, sending the drilling cylinder negative.

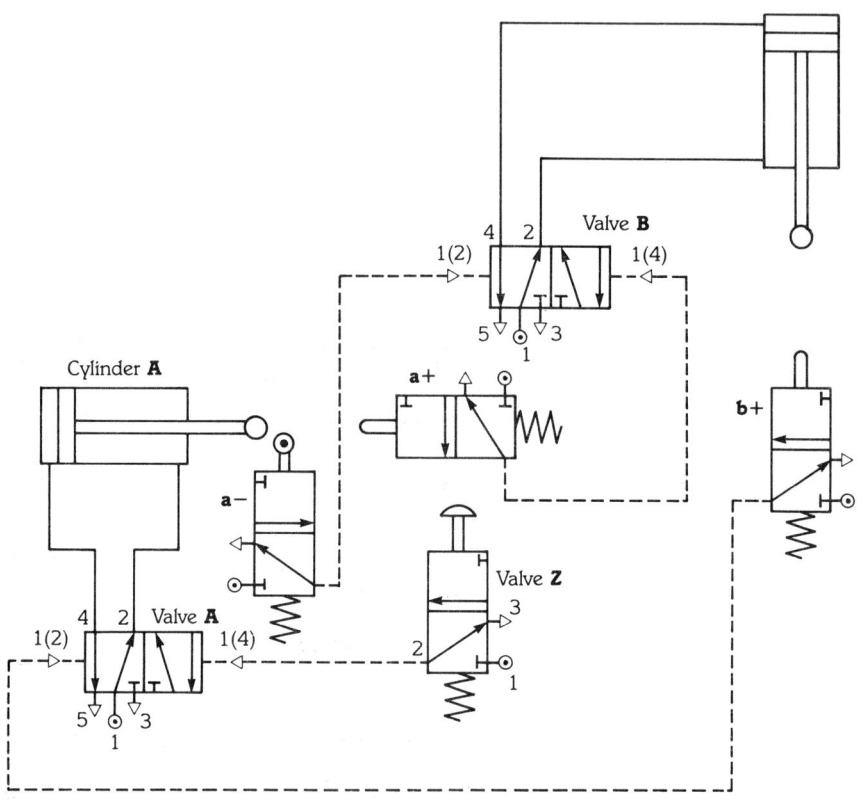

The complete circuit is shown below.

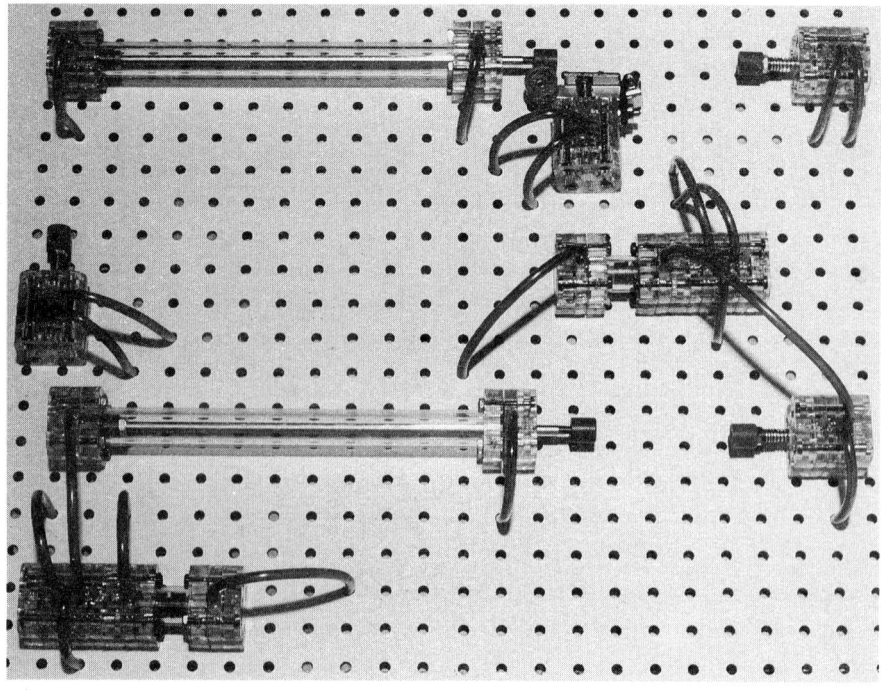

ACTIVITY 22.1

Assemble the complete circuit shown on the opposite page.

Modify the circuit using a roller-trip valve so that it will run continuously once the main air is connected.

ASSIGNMENT 22.1

Design and draw circuits that will operate in the following sequences:

1) A–	2) B+	3) A+
B+	A+	B–
A+	B–	A–
B–	A–	B+

Remember that the last movement in the sequence triggers the first movement when the sequence runs continuously.

Design problem

The people in the dispatch department of a factory want to print a small advertisement on every package that goes out of the factory. The message will be printed by a rubber stamp system. They want to use a simple pneumatic machine that will stamp packages of different sizes. The packages are not spaced regularly on the delivery belt.

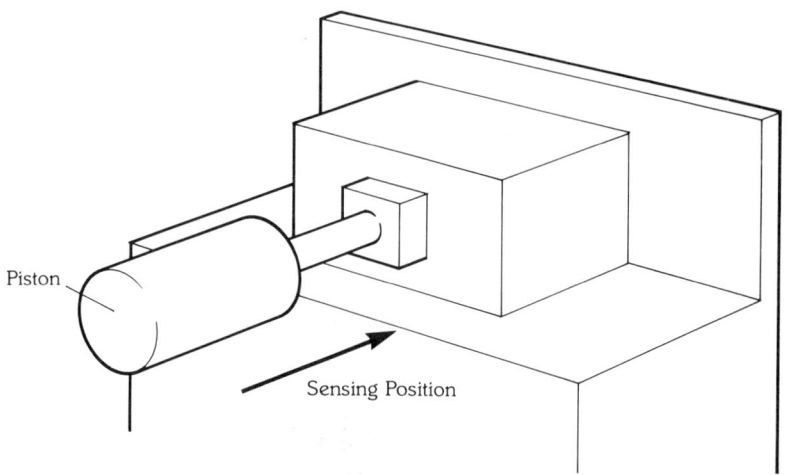

Piston

Sensing Position

Design brief

Design a circuit that will automatically push a printing head on to the front of a package. The printing operation should only occur once when a package is put in position.

Analysis/Research

The sequence of operations for the machine must be:

a) A package arrives in position.
b) The presence of a package is detected.
c) The piston is sent positive.
d) The advertisement is printed on the package.
e) The completion of the printing operation is detected.
f) The piston is sent negative.
g) The package is removed.

There are several problems that must be investigated before a final solution can be proposed.

Design solutions

First problem: How to control the printing mechanism

The printing can be completed by forcing the stamp against the front of the package. This can be done by placing it on the end of the piston of a double-acting cylinder. The cylinder should be remotely operated and be controlled by a double pressure-operated five-port valve.

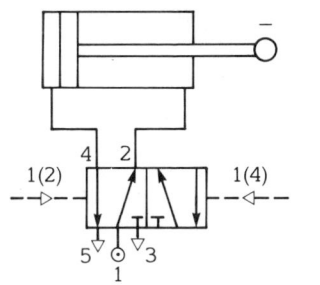

Second problem: How to detect when a package is correctly in place and then send the piston positive

Using an air bleed to detect the presence of a package is a possibility:

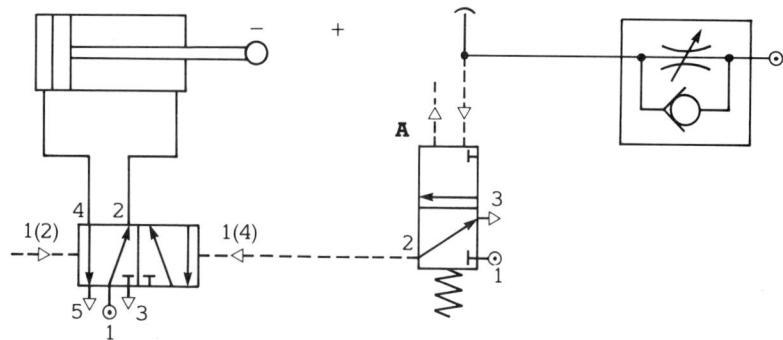

Evaluation

This solution will cause the piston to be sent positive when the air bleed is obstructed. The printing operation may not happen if thin packages fail to block the air jet sufficiently. The circuit needs to be refined.

Using a **long-distance gap-sensing system** any passing package could be detected.

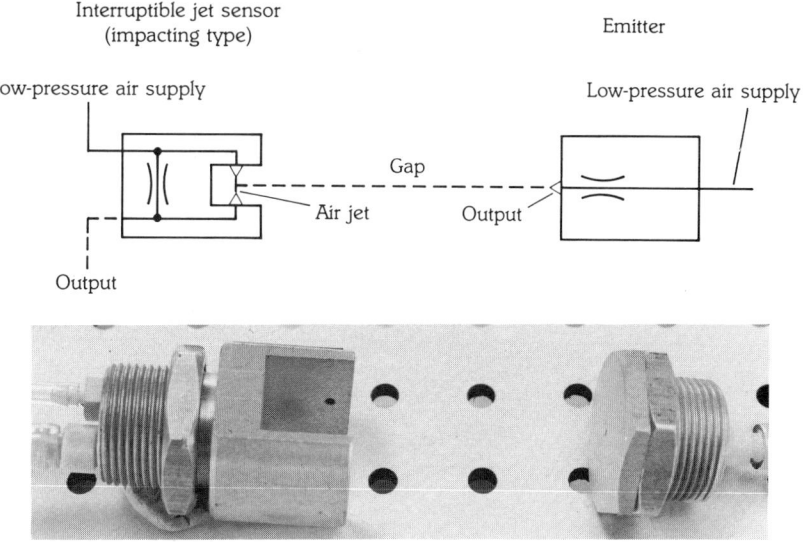

Long-distance gap sensor

The sensing system is completed by the addition of two pressure regulators and gauges. The pressure is set to about one bar. The **emitter** and **sensor** must be carefully aligned (see overleaf).

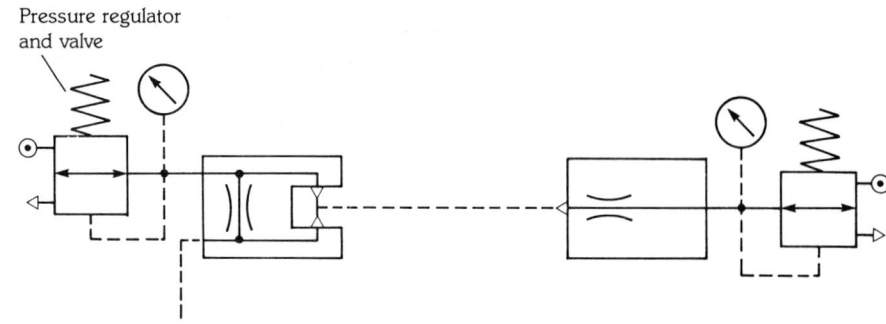

Pressure regulator
and valve

The **interruptible jet sensor** has a jet of low pressure air that
continuously passes across a small gap. If there is any air disturbance
of the sensing air jet (for example from another small air jet) there is
no output from this valve.

The output from the sensor is at a very low pressure. It cannot
operate normal pressure-sensitive valves. A **Boostermite amplifier
valve** will work at these very low pressures. A small signal switches
the state of the valve.

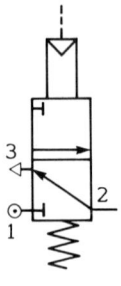

Symbol

Boostermite amplifying valve

The Boostermite valve is connected to the five-port valve as shown
below.

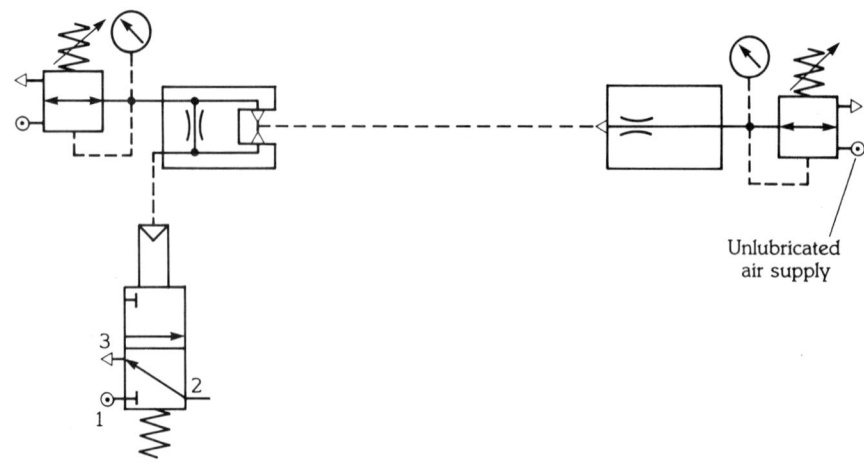

Unlubricated
air supply

Adding this section to the existing circuit produces the following circuit diagram:

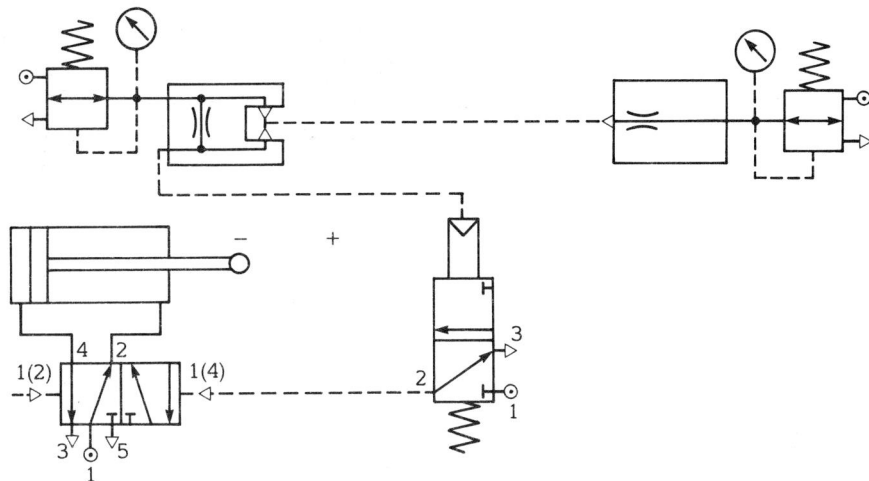

Third problem: How to detect the completion of the printing operation and send the piston negative

The packages may not all be the same size. How can the piston be sent negative when it has stamped the package?

To sense a *reduction* in pressure a pressure-sensitive three-port valve can be used in the circuit. It needs to be connected as a NOT gate. (See Chapter 11.)

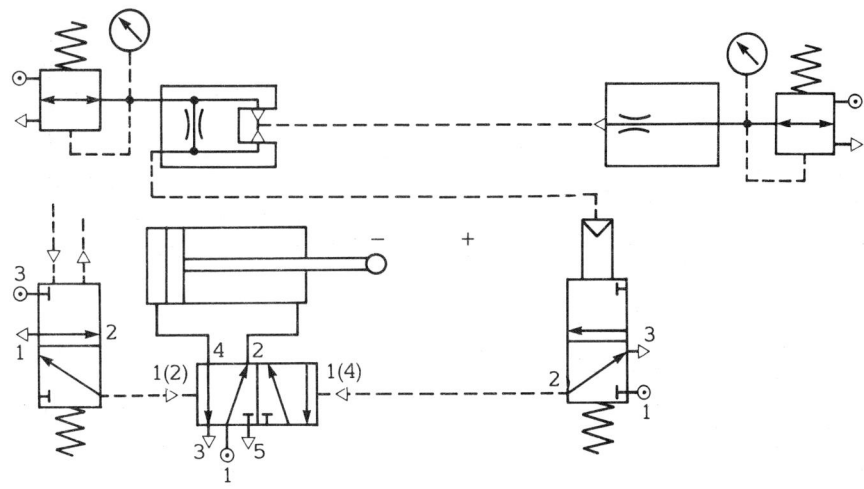

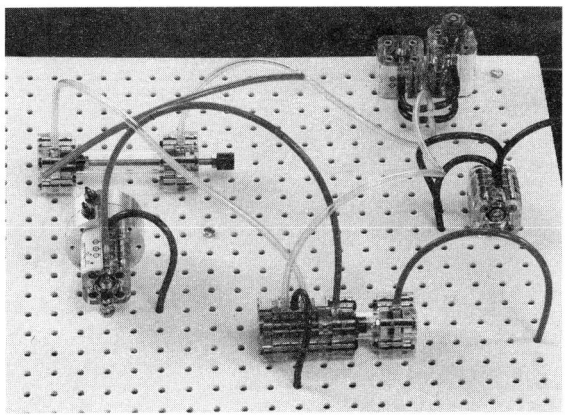

The pressure-sensitive three-port valve plays a crucial part in the detection of a change in pressure.

In this situation the pressure-sensitive three-port valve is used to sense a fall in pressure in the exhaust air as the piston reaches the end of its positive stroke. To make the valve react correctly a flow regulator is added to the exhaust line.

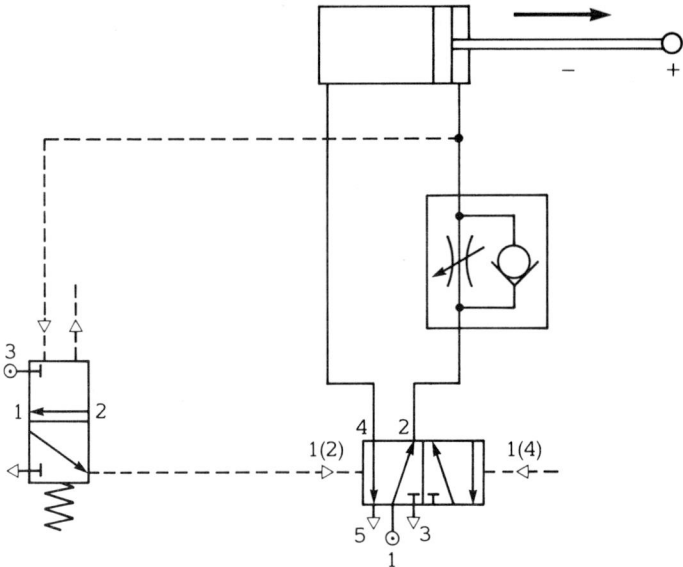

This keeps the air pressure high until the piston has stopped moving. When the air pressure drops the valve changes state, switching the five-port valve. This sends the piston negative.

By sensing **pressure decay** the circuit becomes totally automatic.

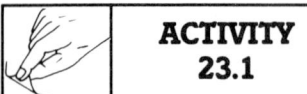

Assemble a model of the stamping machine from materials available in your work area. Construct the pneumatic circuit and built it into the model. Test it and evaluate it.

Using current catalogues from the suppliers of pneumatic equipment,

1) Cost the circuit used in Activity 23.1.
2) Investigate the range of other low-pressure sensors. Make notes on their operation and where they should be used.

Design alternative low-pressure sensing systems that could help solve the design briefs used earlier in this book.

Environmental issues

General issues

Differences in air pressure are a natural feature of the earth's climate.
Low-pressure air movements can result in rustling of leaves or the
uprooting of trees. This source of energy is irregular, but widely
available. It has been used since ancient times. Ships, musical
instruments, windmills, pumps and catapults have been powered by it
(with varying degrees of success!).

The Chinese and Japanese may have been using windmills as early as
2000 BC. Hero of Alexandria wrote a book called *Pneumatica* in the
first century AD. In his book Hero described many air-powered
machines including a wind powered 'Anemurion'. This was a device
like a church pipe organ with compressed air supplied by a wind-
powered pump.

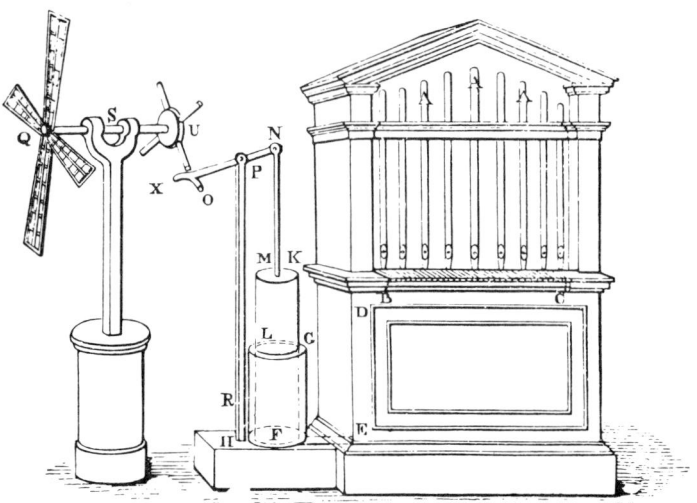

The Anemurion

The use of naturally compressed air as a modern source of energy may be seen in the construction of wind-powered generators. These have been built where there is a regular strong wind. The Orkney Islands, off the Scottish coast, have been used for this purpose.

Wind generator in the Orkneys

The construction of large wind generators standing on tall concrete supports has produced a landscape that is the modern equivalent of the windmills used for draining the low-lying marshlands of East Anglia or Holland during the last century.

Do we want our future countryside and rooftops covered with wind generators?

If not, what are the alternatives?

Talking point

Discuss these questions.

Specific environmental issues

To do work of any sort we need energy. The energy for the 'human machine' is stored in food.

This food is 'burnt' to produce energy. When we pump up a bicycle tyre, some of the energy used in pumping generates heat, but most of it is stored as compressed air in the tyre.

The power needed to compress the large volumes of air used in pneumatic machines has to be supplied from a high-energy source, for example petrol, diesel, electricity or steam.

A petrol- or diesel-driven compressor produces **atmospheric pollution**. Some of the poisonous exhaust gases cause acid rain which damages trees and wildlife.

Even when a compressor is driven by an electric motor pollution is produced. The power station producing the electricity probably runs on fossil fuels. Both fossil fuels and nuclear fuel present hazards that can have damaging effects on the environment.

What about the cost of the **noise pollution** on the environment – the damage to operators' hearing or nervous systems? (Vibration white finger is a nervous system ailment suffered by operators of pneumatic drills.) What about the disruption to other people's work caused by the noise?

What else should we consider?

What are the other hidden costs of producing compressed air, for example the long-term effects on the health of people who use it?

 **ASSIGNMENT 24.1**

Make a list of the environmental consequences that result from using pneumatic systems as a source of control and movement.

Draw an illustration or cartoon that will help to communicate some of these consequences to other pupils.

Calculations

Mathematics are an important aid in solving design problems.

In Chapter 4 units of pressure were mentioned. They are the **bar** and the **pascal**. One pascal is one **newton per square metre**. For our purposes we need a much smaller unit of area. Either square centimetre or square millemitre would be suitable.

Equations

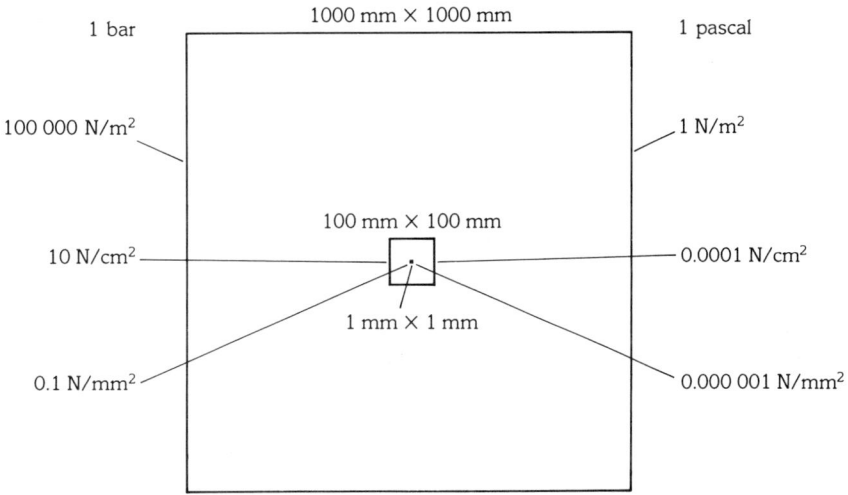

The main equation that is used in pneumatic calculations is:

Force = pressure × area

This is normally written as:

$$F = P \times A$$

There are three ways of arranging this equation. They are:

$$F = P \times A \qquad P = \frac{F}{A} \qquad A = \frac{F}{P}$$

In this book the units used are:

Force: newtons
Pressure: bar
Area: millimetre

In Chapter 9 we investigated the problem of opening a sliding door with a single-acting cylinder. In a real-life situation the designers would have to consider the forces involved in moving the door. They would use this basic equation to help them understand the problem. Say a force of 500 N is needed to open the sliding door and the diameter of the piston is 50 mm. Calculate:

a) The working area of the piston
b) The air pressure required to open the door.

Give your answer in N/mm².

To find the area of the piston:

$$
\begin{aligned}
\text{Area of a circle} &= \pi \times \text{radius}^2 \\
&= 3.14 \times 25 \times 25 \\
&= 1962.5 \text{ mm}^2
\end{aligned}
$$

To find the air pressure required to open the door, select the correct form of the equation:

$$
F = P \times A \qquad P = \frac{F}{A} \qquad A = \frac{F}{P}
$$

$$
\begin{aligned}
P &= \frac{F}{A} \\
&= \frac{500 \text{ N}}{1962.5 \text{ mm}^2} \\
&= 0.255 \text{ N/mm}^2 \text{ (or 2.55 bar or 255 000 Pa)}
\end{aligned}
$$

Note: To convert N/mm² to bars multiply by 10.

Summary of useful equations

$$
\text{Force} = \text{pressure} \times \text{area}
$$

$$
\text{Assume } \pi = \frac{22}{7}
$$

$$
\text{Area of a circle} = \pi \times \text{radius}^2
$$

To calculate the efficiency of a system you will need this equation:

$$
\text{Work done} = \text{force} \times \text{distance moved by the force}
$$

 **ASSIGNMENT 25.1**

You need a cylinder that will open a sliding door. The air pressure available is 0.255 N/mm². A force of 3178 N is required to move the door. Calculate the area of piston that is needed.

Using suppliers' catalogues find out the cost of a suitable cylinder. (Assume that the door has to be opened 550 mm.)

 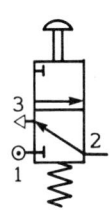

Name: Push-button operated spring-return three-port valve.

Applications: Manual or hand operation *only*, do not use for cylinder operation.

 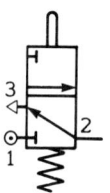

Name: Plunger-operated spring-return three-port valve.

Applications: Cylinder operation or hand operation.

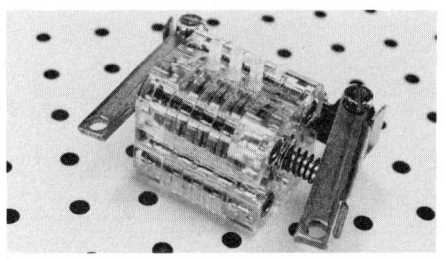 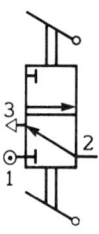

Name: Lever set/reset three-port valve.

Applications: Manual operation.

 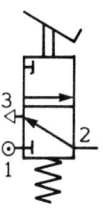

Name: Foot-operated lever spring-return three-port valve.

Applications: Foot operation.

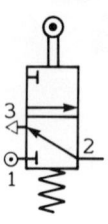

Name: Roller-trip-operated spring-return three-port valve.

Applications: Operated by piston contact.

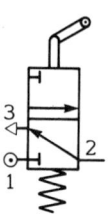

Name: Lever set/reset spring-return three-port valve.

Applications: Operated by one direction of piston movement.

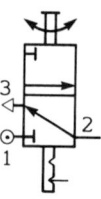

Name: Key-lock-operated three-port valve.

Applications: Security purposes, secured on/off valves.

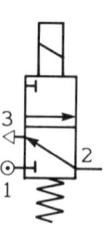

Name: Solenoid-operated spring-return three-port valve.

Applications: Remote operation of pneumatic circuit by an electronic switch.

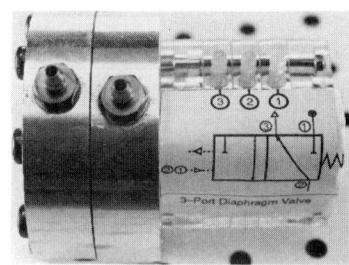 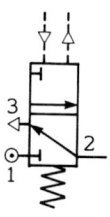

Name: Pressure-sensitive diaphragm-operated spring-return three-port valve.

Applications: Low-pressure operation (0.5 bar), either pressure or vacuum, depending on input port.

 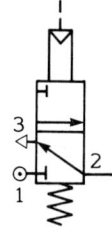

Name: Low-pressure-operated amplifier spring-return three-port valve, or Boostermite valve.

Applications: Very low-pressure sensing. Maximum pressure must not be more than 0.3 bar. Operates on 'very low-pressure' signal to control 'normal' pressure.

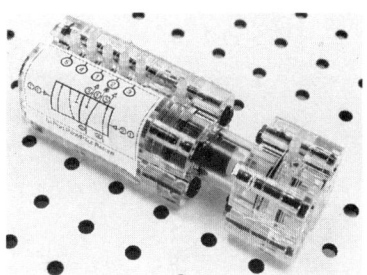 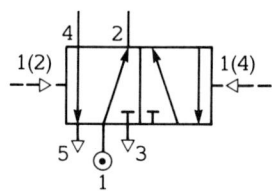

Name: Double pressure-operated five-port valve.

Applications: Control of double-acting cylinder, operated by air pressure signals.

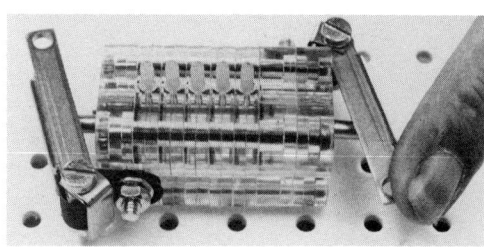 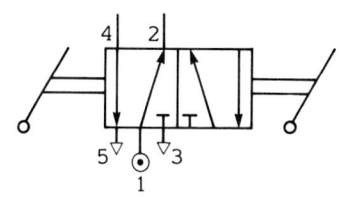

Name: Lever set/reset five-port valve.

Applications: Manual control.

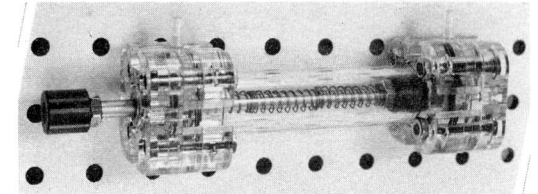 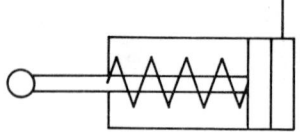

Name: Single-acting cylinder.

Applications: Cylinder powered in one direction only. Can be made with spring producing either positive or negative stroke.

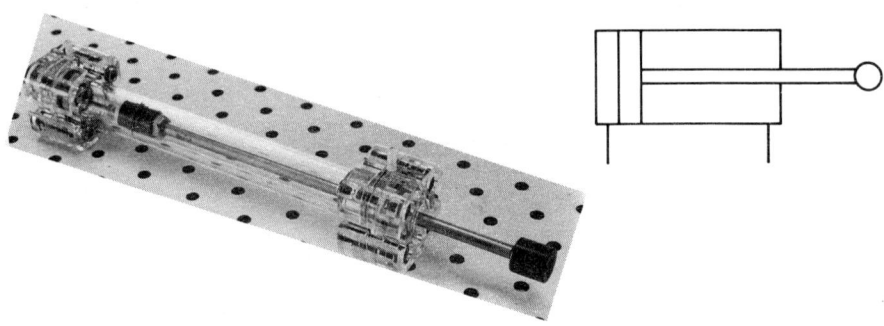

Name: Double-acting cylinder.

Applications: Cylinder powered in both directions. Greater power available on both the positive and negative strokes.

  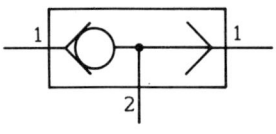

Name: Shuttle valve.

Applications: OR gate operations.

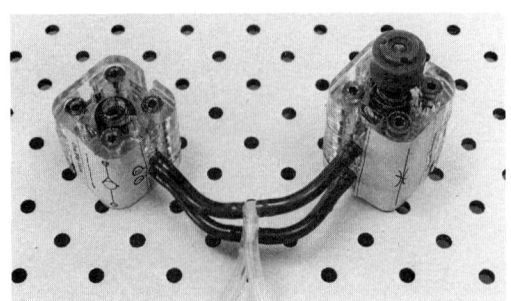 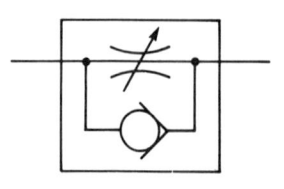

Name: Flow restrictor (flow regulator).

Applications: Cylinder speed control. Time delay circuits.

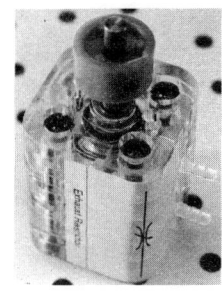 

Name: Throttle valve.

Applications: Cylinder speed control. Time delay circuits.

Name: Reservoir.

Applications: Time delay circuits.

 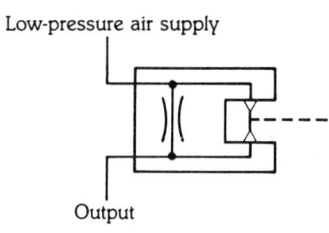

Name: Interruptible jet sensor, impacting type.

Applications: Remote sensing of objects.

 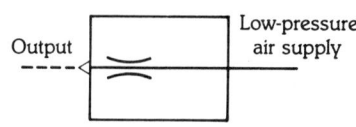

Name: Emitter.

Applications: Remote sensing of objects.

 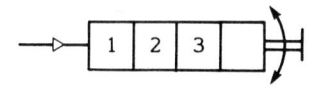

Name: Pneumatic counter.

Applications: Counting air pressure pulses.

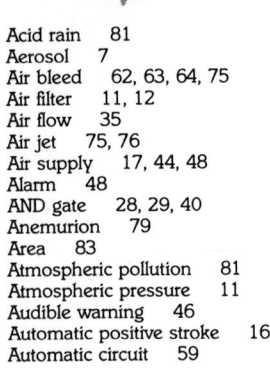

Index